PLANT BIOTECHNOLOGY

PLANT BIOTECHNOLOGY

C.B. Nirmala
Head
Department of Biotechnology
SRM Arts and Science College, Chennai

G. Rajalakshmi
Lecturer
Department of Biotechnology
DG Vaishnav College, Chennai

Chandra Karthik
Researcher
Indiana University-Purdue University
Indianapolis, USA

MJP PUBLISHERS

MJP PUBLISHERS

© Publishers, 2024   47, Nallathambi Street
All rights reserved   Triplicane
Chennai 600 005

Publisher : J.C. Pillai

To

Our master,
the treasury of knowledge
and doer of actions

PREFACE

Biotechnology has brought in a major revolution in science in the last three decades. It has contributed many technologies that have tremendous applications in the fields of medicine, agriculture, horticulture, aquaculture, textiles, tanneries, baking and brewing industries.

The book initially deals with basic information about the biological processes occurring in plants. It explains the manipulation of these processes by molecular methods to produce technologies in favour of mankind. Finally, it discusses about the achievements of Biotechnology in plant sciences.

The book aims to present the complicated concepts of Biotechnology in simple terms with the help of neat diagrams so as to provide a good understanding of the subject to beginners, students and entrepreneurs.

C. B. Nirmala
G. Rajalakshmi
Chandra Karthik

ACKNOWLEDEMENTS

I express my heartfelt thanks to the Almighty for His blessings.

I thank my mentor, Prof. V.S. Sundaralingam, for his invaluable help during my academics and career.

My sincere thanks to Prof. A. Balasubramanian, former Principal, K.C.S. Kasi Nadar College of Arts and Science, Chennai, who not only guided my doctoral work but also taught me perfection in writing and editing research output.

I extend my thanks to Dr.George Thomas, General Manager, Interfield Laboratory, Cochin and Dr. S.P. Palaniappan, Manager, MphasiS (Tidel Park), Chennai, who trained me in Molecular Biology and Microbiology respectively.

I thank my husband, Dr.V. Mohankumar, Global International, Chennai, for his continuous encouragement and support.

I thank my family members and especially my brother, Mr. S.Murugan, and his family members, Aruna, Manohari and Nishanth for their help.

I thank Mrs.R. Alfana for typing the hand written script of the book.

I thank Dr. S. Ragupathy, University of Guelph, Dr. A. Premkumar University of Stockholm and Dr. P. Loganathan, North Carolina University, for their valuable contributions.

I am amazed by the editorial group of MJP Publishers for their perfection in correction and for their hospitality throughout this project work. I sincerely thank each and every one of them.

C.B. Nirmala

CONTENTS

PART I

PLANT GENOME ORGANIZATION

Nuclear Genome

Chloroplast and its Genome

Mitochondrion and its Genome

Transposable Elements

The term cell was coined by Robert Hooke (1665), to describe a hollow space in the thin slice of cork that he saw under the self-built compound microscope. Two German scientists, M.J. Schleiden and T.S. Schwann independently proposed the **cell theory** in the year 1839 that the tissues of living beings, plants and animals are all made up of cells.

According to the organization of the cellular components, cells can be divided into two types, (i) **prokaryotic** and (ii) **eukaryotic** cells. In prokaryotic cells, membrane-bound cytoplasmic organelles like mitochondria, Golgi bodies, endoplasmic reticulum and nucleus are not present. Some examples of prokaryotes are bacteria, blue-green algae and other lower organisms. The eukaryotic cells have membrane-bound organelles such as true nucleus, mitochondria and many other highly specialized organelles carrying out specific functions. All the cells of higher organisms belong to this group.

Despite their differences, cells of all kinds share certain structural features. The cell has an outer limiting **plasma membrane**, an internal volume of soluble, insoluble and suspended substances, the **cytoplasm** and a thicker nucleic-acid-containing area.

Mitochondria are specialized organelles in the cell, surrounded by two membranes. In their inner membrane, **cytochromes** are present that are involved in the electron transport chain reaction and in oxidative phosphorylation where the energy molecule, adenosine triphosphate (ATP) is synthesized. Chloroplasts are vital organelles where the plant-food is synthesized. Pigment molecules (chlorophylls) in the chloroplast absorb light energy from the sun and are used to reduce CO_2 to form carbohydrates such as starch and sucrose. It can also synthesize ATP.

Among the cytoplasmic organelles, mitochondria and chloroplast are discussed in detail in this chapter as biosynthesis of many of the essential compounds occurs in these organelles.

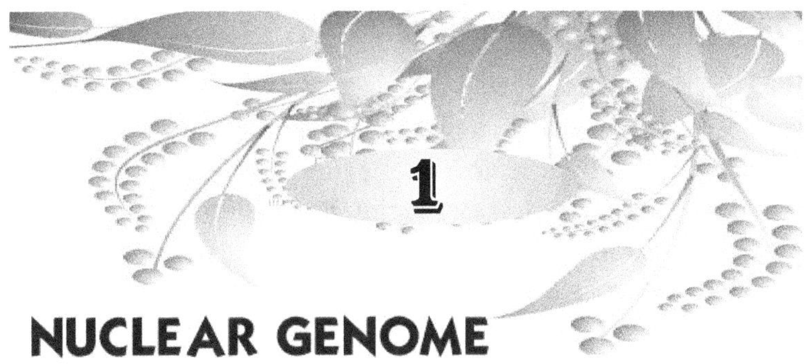

NUCLEAR GENOME

Deoxyribonucleic acid (DNA) has the genetic information that specifies the growth, form and developmental pattern of all organisms, both plants and animals.

NUCLEUS AND CHROMATIN ORGANIZATION

Generally in a plant cell, DNA is contained in the nucleus that is usually spherical. Mostly the nuclei of higher plants are 3–20 µm across but the giant nuclei of the alga *Acetabularia* measures up to 150 µm. The nucleus is surrounded by two membranes that are separated by a perinuclear space and perforated by pores that range from 50 to 100 nm in size. The nuclear membrane is connected to the endoplasmic reticulum and is sometimes in close association with the membranes of mitochondria and chloroplasts. During nuclear division, this association is temporarily dismantled while the chromosomes segregate.

A major nuclear organelle called the nucleolus arises after mitosis at specific chromosomal locations called nucleolar organizers. Nucleoli develop into prominent organelles without a limiting membrane. They contain DNA, and fibrils and granules of RNA and protein. They are the sites of transcription of the rRNA genes, and of processing and partial assembly of 80S ribosomes destined for the plant cytosol.

Chromatin has a highly complex structure with several levels of organization. The simplest level is the double-helical structure of DNA. At a more complex level, the DNA molecule is associated with proteins and is highly folded to produce a chromosome. The DNA is first aggregated into nucleosomes (Figure 1.1) and then wound to form the 30-nm chromatin fibres. Further coiling leads to the condensed metaphase chromosome.

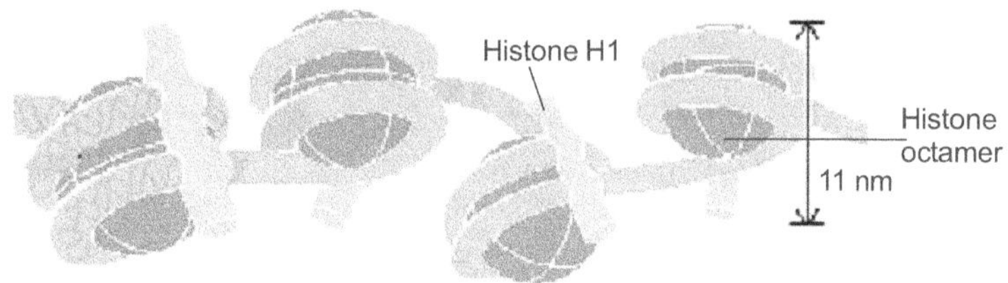

Figure 1.1 Nucleosomes

The nucleosome is a core particle consisting of DNA wrapped about two times around an octamer of eight histone proteins (two copies each of H2A, H2B, H3, and H4), much like the thread wound around a spool. (The fifth type of histone, H1, is not a part of the core particle but plays an important role in the nucleosome structure). Nucleosome is the primary structural unit of chromatin giving a beads-on-a-string appearance to the DNA. With this main idea let us move on to further discussion in the following sections.

PACKAGING OF DNA

DNA is incredibly long and yet it is packed within a nucleus of diameter only a few tens of microns. As a result, the DNA must be intensively packed; but at the same time it must be extensively available.

DNA is wound (two times) around a core complex of four small proteins (H2A, H2B, H3 and H4) termed histones (approximately 180 bp of DNA is wound twice around the structure). These form a barrel-shaped core octamer structure called the nucleosome as discussed in the earlier section. The compacting effect of the nucleosome reduces the length of the DNA by a factor of 6 (Figure 1.2). These nucleosomes fold on themselves to form a dense, tightly packed structure.

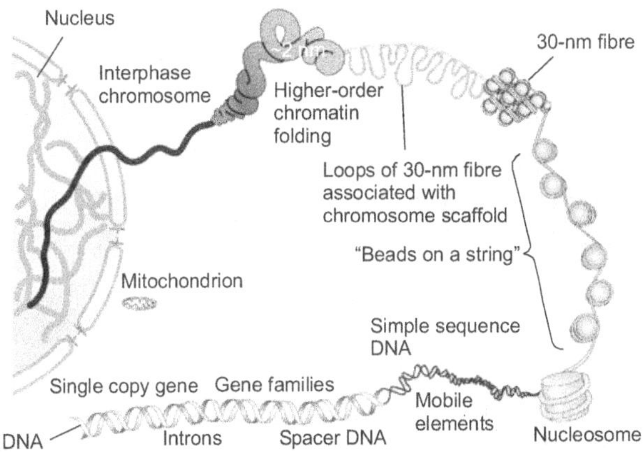

Figure 1.2   Chromatin condensation

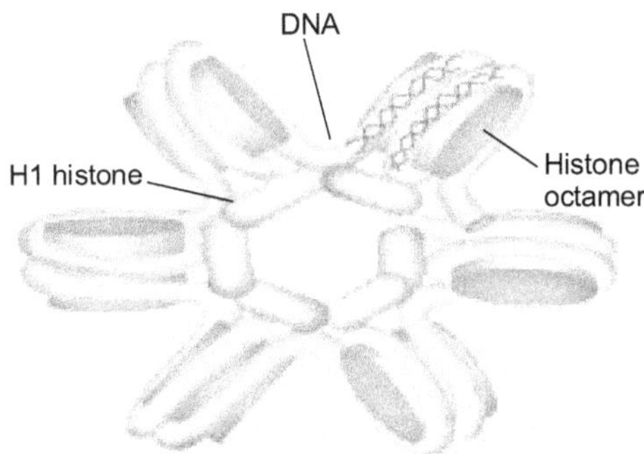

Figure 1.3   Model of chromatin condensation into 30-nm fibre. View along the axis of one turn of solenoid.

Nucleosomes are arranged in tiers resulting in 30-nm chromatin fibre, thus further reducing the length of the DNA by a factor of 7. Each tier is called a solenoid comprising of six nucleosomes (Figure 1.3).

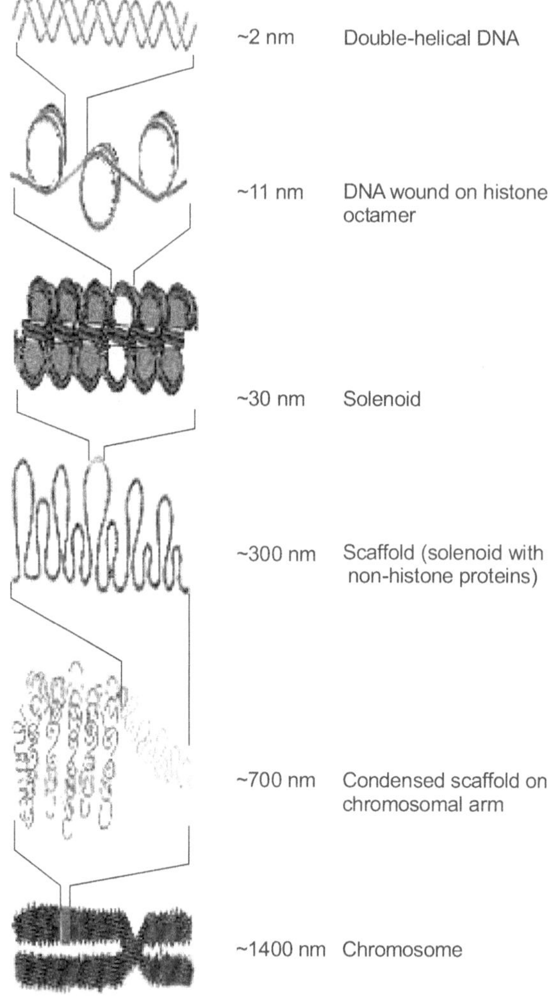

Figure 1.4   Various orders of folding of DNA molecule packaged into a chromosome

Fibres consisting of compactly arranged solenoids form loops that are anchored to the nuclear matrix by scaffolding proteins. These structures are further folded and looped through the interaction with other non-histone proteins that ultimately form chromosome structures.

The DNA is attached at certain positions within the scaffold, usually coinciding with the origins of replication.

Many other DNA-binding proteins are also present, such as high mobility group (HMG) proteins, which assist in promoting certain DNA conformations during processes such as replication and active gene expression. The dimension of various orders of folding is shown in Figure 1.4.

DNA has the genetic information for replication and protein synthesis. All cells undergo protein synthesis continuously in order to maintain metabolism and life. But replication is restricted to some cells. DNA function as a template for both replication and protein synthesis have been discussed further in detail.

DNA REPLICATION

Most actively dividing plant cells complete a cycle of growth and cell division in 15–40 hours depending on the species and temperature. DNA replication and histone synthesis, which last for 7–11 hours are confined to part of the interphase cycle known as the S phase.

Replication of DNA proceeds in both directions from each origin of replication. Each segment of DNA thus synthesized is called a replicon. These replicons are 20–30 or 60–90 kb long. The replication forks that number from 5000 to 60,000 per diploid genome move at about 10 bp per hour. There are 2–25 families of replicons in different higher plants. Each replicon in a family undergoes DNA synthesis at the same time during the S phase but separate families are active at different times.

DNA synthesis is initiated at the replication fork by nicking and unwinding of the DNA by a topoisomerase (unwinding enzyme or helicase). The separated DNA strands are bounded by the enzyme DNA polymerase. DNA replication is a semi-conservative process with complementary strands being synthesized in the 5′ to 3′ direction by DNA polymerase, using the separated strands of the helix as template. It is not possible for both the new strands to be synthesized continuously because the enzyme polymerizes nucleotides in the 5′ to 3′ direction only.

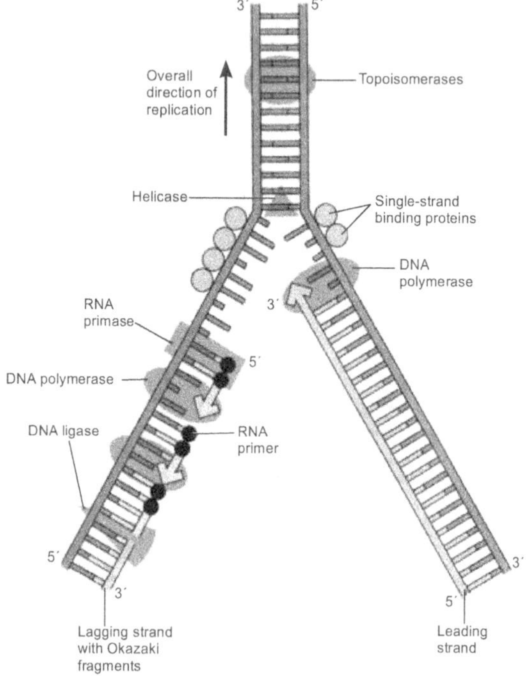

Figure 1.5 DNA replication

Therefore, one leading strand is synthesized in the 5′ to 3′ direction and a lagging strand again elongates in the 5′ to 3′ direction by the ligation of short fragments of DNA called Okazaki fragments (Figure 1.5).

Short stretches of RNA sequences are synthesized at the origin of fork. These RNA fragments and the enzyme RNA polymerase are used to prime the continuous synthesis of the leading and short stretches of DNA in the lagging strand. The Okazaki fragments are about 200 nucleotides long. RNA primers are then removed from the Okazaki fragments by a nuclease. The resulting gaps in the DNA are repaired by DNA polymerase II (DNA Pol II). The adjacent DNA segments are ligated together to produce a new daughter strand.

In the higher plants, after DNA replication, up to 25% of the cytosine residues are methylated by DNA methylase that uses S-adenosylmethionine as the methyl donor.

PROTEIN SYNTHESIS

DNA contains the "genetic template" for proteins. The central dogma of molecular biology indicates that there is a directional flow of genetic information from DNA to protein (Figure 1.6).

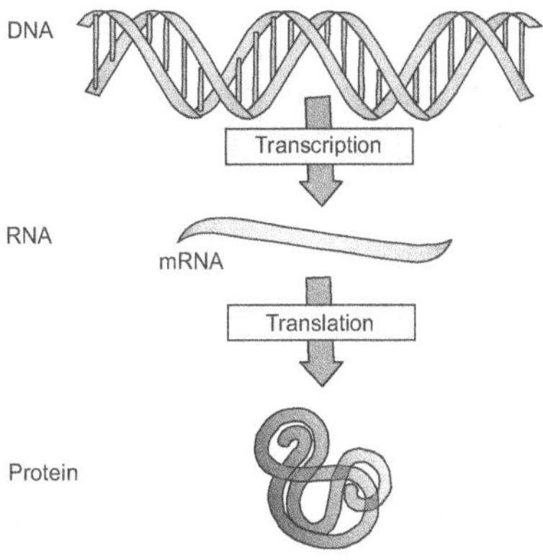

Figure 1.6   The central dogma

DNA is found in the nucleus whereas protein synthesis occurs in the cytoplasm on the ribosomes. So the "genetic information" must be transferred to the cytoplasm where proteins are synthesized. Gene expression includes transcription and translation. Transcription is the process where DNA, the genetic template for a protein, is copied into RNA (mRNA) and transported out to the cytoplasm. Translation is the process where the RNA template serves as a series of codes for the amino acid sequence thereby synthesizing the protein.

STRUCTURE OF PLANT GENES

The general structure of a plant nuclear gene including its control sequences is shown in Figure 1.7. The region of the chromosome lying 5'-end of the transcribed region is called the promoter region. This promoter contains signals that are important in controlling the expression of the gene. It contains two types of elements, namely distal elements and proximal elements.

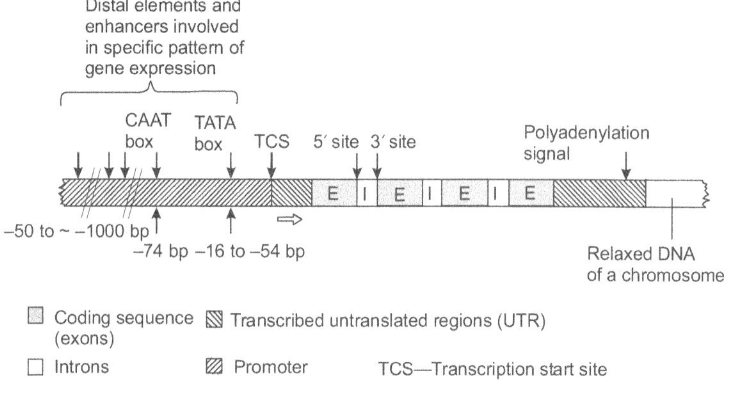

Figure 1.7 Structure of a plant nuclear gene

Distal elements are farthest from the transcription start site, may be several hundred bases away from the proximal elements that are normally situated within 75 bp of the transcription start site (TCS). The distal elements have been studied for a number of genes and are responsible for the particular pattern of expression of a gene. The proximal elements are involved in the general control of transcription, that is, they are found in most genes and consist of 2 motifs: the CAAT or AGGA box and the TATA box. These regulate the level of transcription. The TATA box is closer to the TCS, normally 30 bp 5' of the TCS, (written as –30 bp), and is present in virtually all the expressed plant genes. The TATA box is involved in the orientation of RNA polymerase II, the enzyme responsible for mRNA synthesis (transcription).

The promoter is followed by the sequence of the primary transcript. This sequence contains signals important for the processing of the primary transcript to produce mRNA.

Enhancers are the sequences that can be situated at a considerable distance from the TCS and act in a position- and orientation-independent manner to stimulate the expression of a gene, e.g. in the soybean storage protein (conglycinin) gene, a sequence of four repeats A(A/G/C)CCA behaves like an enhancer and can give rise to a 25-fold increase in the level of gene expression.

Classes of RNA

RNA molecules perform a variety of functions in the cell. There are three classes of RNA. They are

   i. Ribosomal RNA (rRNA),

  ii. Messenger RNA (mRNA) and

 iii. Transfer RNA (tRNA).

Ribosomal RNA, along with ribosomal protein subunits, makes up the ribosome, the site of protein assembly.

Messenger RNA (mRNA) carries the coding instructions for polypeptide chains from DNA to the ribosome. After attaching to a ribosome, an mRNA molecule specifies the sequence of the amino acids in a polypeptide chain and provides a template for joining amino acids. Large precursor molecules, which are termed pre-messenger RNAs (pre-mRNAs), are the immediate products of transcription in the eukaryotic cells. Pre-mRNAs are modified extensively, before they exit the nucleus for translation into protein. Bacterial cells do not possess pre-mRNA; in these cells, transcription takes place concurrently with translation.

Transfer RNA (tRNA) serves as a link (adaptor molecule in protein synthesis) between the coding sequence of nucleotides in the mRNA and the amino acid sequence of a polypeptide chain. Each tRNA attaches to one particular type of amino acid and helps to incorporate that amino acid into the polypeptide chain.

Additional classes of RNA molecules are found in the nuclei of eukaryotic cells. Small nuclear RNAs (snRNAs) combine with small nuclear protein subunits to form small nuclear ribonucleoproteins (snRNPs, affectionately known as "snurps"). The snRNPs are analogous to ribosomes in structure, only smaller, and they typically contain a single RNA molecule combined with approximately 10 small nuclear protein subunits. Some snRNAs participate in the processing of RNA, converting pre-mRNA into mRNA and small nucleolar RNAs (snoRNAs) that take part in the processing of rRNA.

RNA polymerase　Eukaryotic cells contain three RNA polymerases: RNA polymerase I, which transcribes rRNA; RNA polymerase II, which transcribes pre-mRNA and some snRNAs; and RNA polymerase III, which transcribes tRNAs, small rRNA, and some snRNAs.

TRANSCRIPTION

It is the first step (Figure 1.8) in the transfer of genetic information from the genotype to the phenotype. RNA, like DNA, is a polymer consisting of nucleotides joined together by phosphodiester bonds. However, there are several important differences in the structures of DNA and RNA (Table 1.1).

Table 1.1 Comparison of DNA and RNA

Characteristics	DNA	RNA
Composed of nucleotides	Yes	Yes
Type of sugar	Deoxyribose	Ribose
Presence of 2′-OH group	No	Yes
Bases	A, G, C, T	A, G, C, U
Nucleotides joined by phosphodiester bonds	Yes	Yes
Double- or single-stranded	Usually double	Usually single
Secondary structure	Double helix	Many types
Stability	Quite stable	Easily degraded

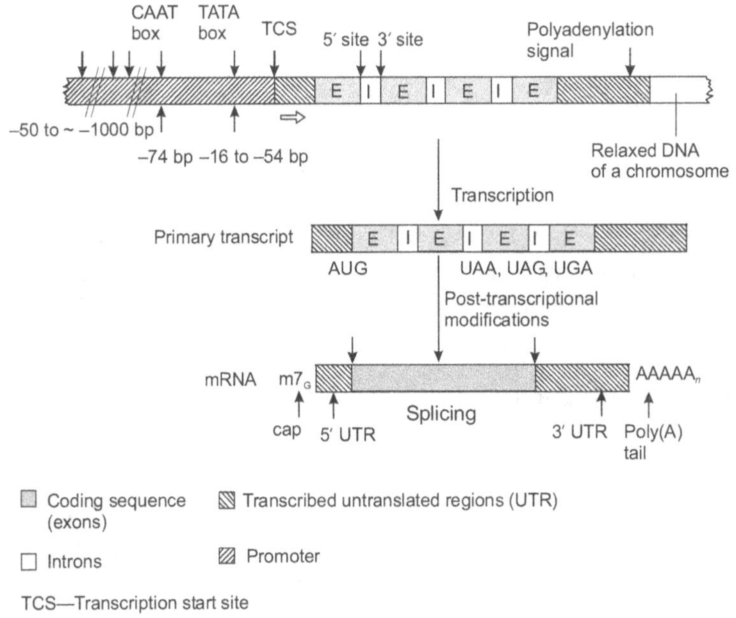

Figure 1.8 Process of transcription

The Transcription Unit

A **transcription unit** is a stretch of DNA that codes for an mRNA molecule. Included within a transcription unit are three critical regions: a promoter, coding sequence, and a terminator (Figure 1.9).

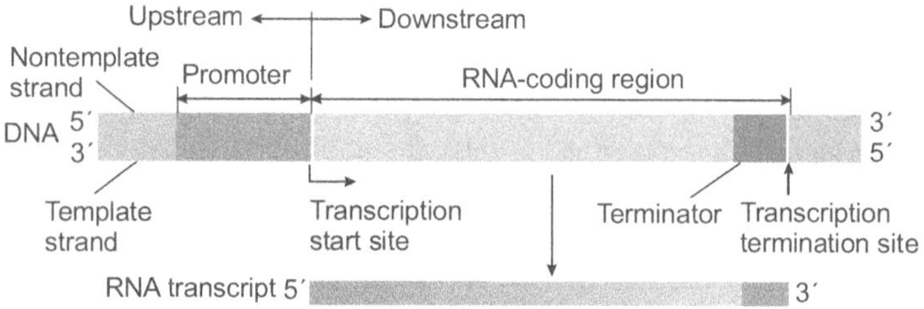

Figure 1.9 A transcription unit

The **promoter** is a DNA sequence that the transcription apparatus recognizes and binds. It indicates which of the two DNA strands is to be read as the template and the direction of transcription. The promoter also determines the transcription start site, the first nucleotide that will be transcribed into RNA. In most of the transcription units, the promoter is located next to the transcription start site but is not itself transcribed.

The second critical region of the transcription unit is the **RNA-coding region,** a sequence of DNA nucleotides that is copied into an RNA molecule. The third component of the transcription unit is the **terminator,** a sequence of nucleotides that signals where transcription is to end. Terminators are usually part of the coding sequence, that is, transcription stops only after the terminator has been copied into RNA.

Molecular biologists often use the terms **upstream** and **downstream** to refer to the direction of transcription and the location of nucleotide sequences surrounding the RNA coding sequence. The transcription apparatus is said to move downstream during transcription—it binds to the promoter (which is usually upstream of the start site) and moves towards the terminator (which is downstream of the start site).

By convention, the sequence on the non-template strand is written with the 5′-end on the left and the 3′-end on the right. The first nucleotide transcribed (the transcription start site) is numbered +1; nucleotides downstream of the start site are assigned positive numbers, and nucleotides upstream of the start site are assigned negative numbers.

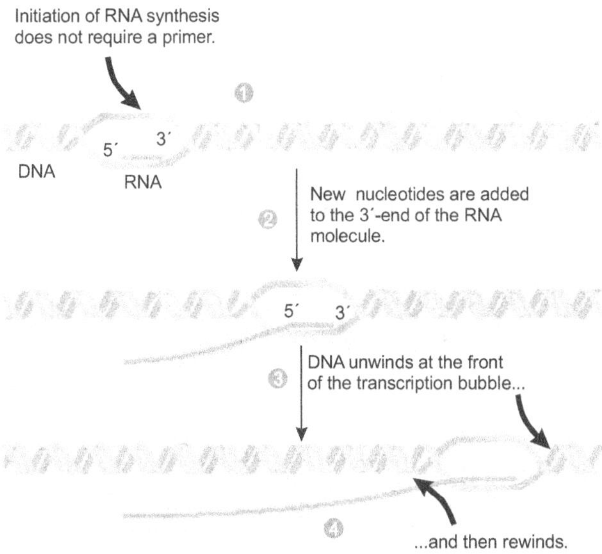

Figure 1.10 DNA as the template for RNA synthesis (transcription)

The process of transcription consists of three stages.

1. Initiation

2. Elongation

3. Termination

Promoters are recognized by the transcription apparatus and are required for transcription. They contain short consensus sequences imbedded within longer stretches of DNA. Transcription begins at the start site, which is determined by the consensus sequences (initiation). A short stretch of DNA is unwound near the start site (Figure 1.10); RNA is synthesized from a single strand of DNA as a template (elongation), and the DNA is rewound at the lagging end of the transcription bubble. Terminators consist of sequences within the RNA coding region; RNA synthesis ceases (termination) after the terminator has been transcribed.

PROCESSING OF mRNA

Following the transcription, the primary transcript is processed in the following three ways and is also referred to as post-transcriptional modification.

1. 5′ capping—addition of the modified base, 7-methyl guanosine, to cap the 5′-end.
2. Splicing—removal of introns.
3. Polyadenylation—addition of a poly(A) tail.

5′ Capping

mRNA capping occurs at the 5′-end of mRNA via addition of a 5′, 5′-PPP linked guanosine (GpppN) (Figure 1.11). Capping occurs co-transcriptionally and is thought to be added before more than ~20 nucleotides are transcribed into RNA and does not normally occur on free RNA. The true substrate for capping is a ternary complex of template DNA, RNA polymerase, and nascent RNA. The 5′ cap is required for (i) mRNA splicing, (ii) export from the nucleus, (iii) stability of the mRNA and (iv) translation.

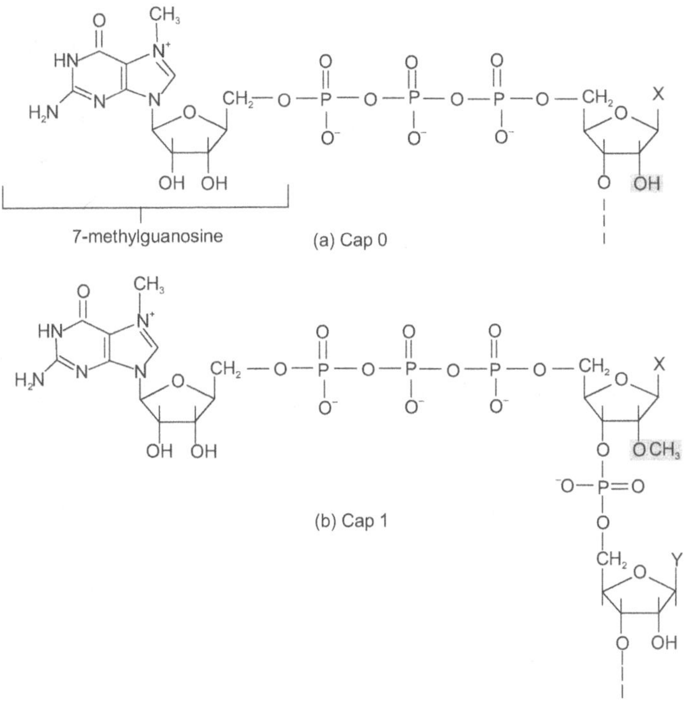

Figure 1.11 mRNA 5' cap (a) Before capping (b) After capping

S-adenosylmethionine (S-Ado-Met) is the source of the methyl (CH_3) group for the two methylation steps. Most eukaryotic mRNAs contain a 5' cap of 7-methylguanosine linked to the 5'-end of the mRNA in a novel 5', 5'-triphosphate linkage. The added guanosine base is not encoded in the DNA template. Another methyl group is added at 2'-O position of the first template-encoded nucleotide.

Cap functions Some of the functions of the cap are the following.

1. *Translation* N7-methylation of the cap is critical to cap function in promoting translation of most eukaryotic cellular mRNAs.

A translation initiation factor binds the mRNA cap and is associated with the recruitment of mRNA to 40S ribosome during translation initiation.

2. *mRNA processing and transport*   The 5′ cap may also target mRNA for processing, polyadenylation and transport.

mRNA Splicing

Many plant nuclear genes contain intervening sequences or introns. These sequences are excised from the primary transcript (pre-mRNA or hnRNA, i.e., heterogeneous nuclear RNA) by a process known as splicing. Splicing proceeds through 2 steps. In the first step, cleavage occurs at the 5′ splice site, with the formation of a covalent attachment of the cut 5′-end of the intron to an internal adenosine located within the intron, upstream of the 3′ splice site (Figure 1.12). In the second step, the 3′ splice site is cleaved and the two exon (coding) sequences are ligated.

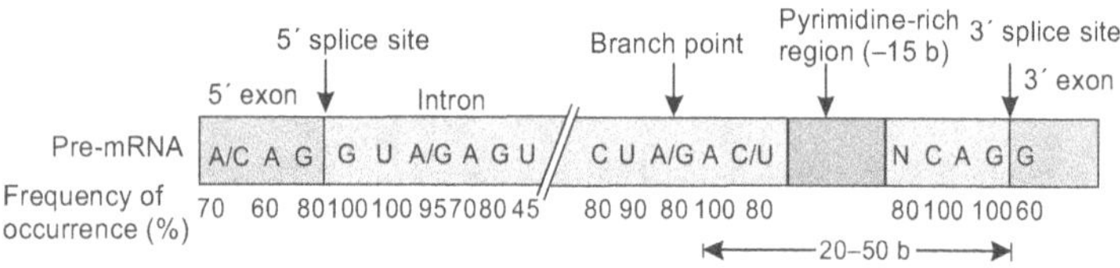

Figure 1.12 Consensus sequences around 5′ and 3′ splice sites in pre-mRNA

Splicing occurs at short, conserved sequences in pre-mRNAs via two transesterification reactions. The location of exon–intron junctions (i.e., splice sites) in a pre-mRNA can be determined by comparing the sequence of genomic DNA with that of the cDNA prepared from the corresponding mRNA. Sequences that are present in the genomic DNA but are absent in the cDNA represent introns and indicate the positions of splice sites. Such analysis of a large number of different mRNAs revealed moderately conserved, short consensus sequences at intron–exon boundaries in eukaryotic pre-mRNAs. The most conserved nucleotides are the (5′) GU and (3′) AG found at the ends of introns (Figure 1.12). Deletion analyses of the centre portion of introns in various pre-mRNAs have shown that generally only 30–40 nucleotides at each end of an intron are necessary for splicing to occur.

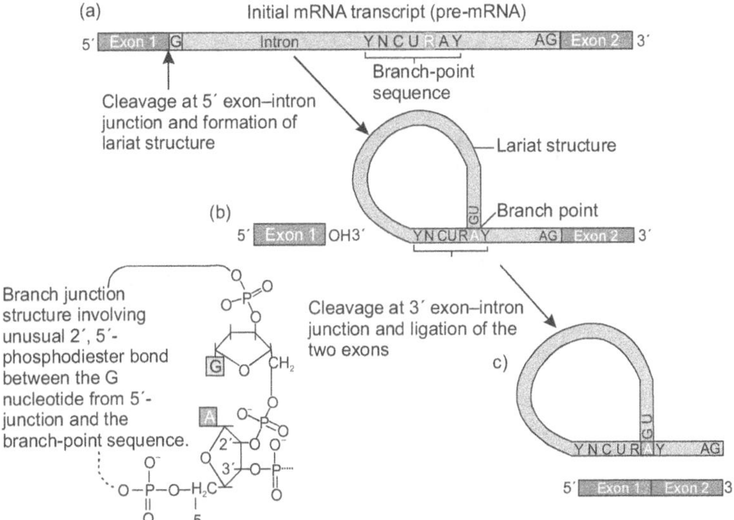

Figure 1.13 Splicing

Studies on splicing of pre-mRNAs *in vitro* led to the conclusion that introns are removed as a lariat structure (⌒) in which the 5′G of the intron is joined in an unusual 2′, 5′-phosphodiester bond to an adenosine (A) near the 3′-end of the intron (Figure 1.13).

This A residue is called the branch point because it forms an RNA branch in the lariat structure. Splicing of exons proceeds via two sequential transesterification reactions. Spliceosomes, assembled from snRNPs and a pre-mRNA, carry out splicing. Small nuclear RNAs (snRNAs) assist in the splicing reaction (Figure 1.14).

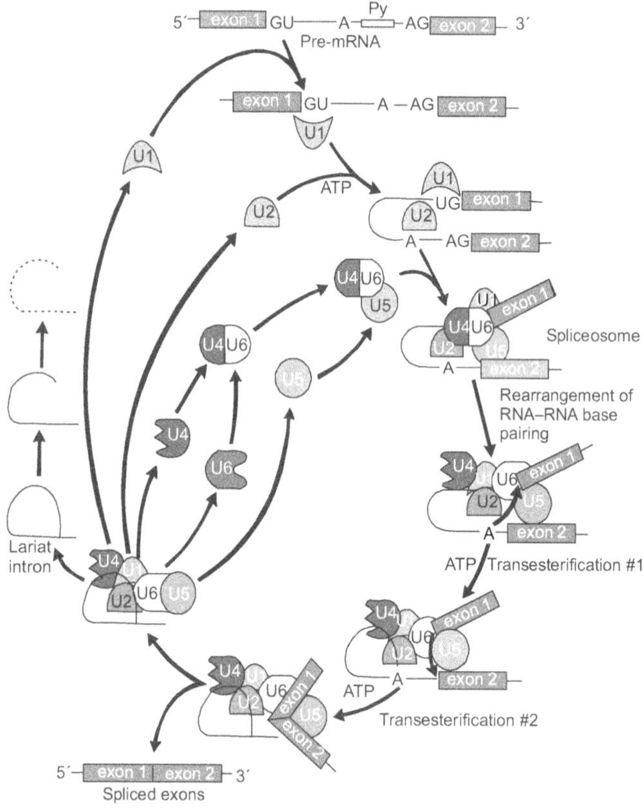

Figure 1.14 The spliceosome assembly and splicing of pre-mRNA

The short consensus sequence at the 5′-end of introns is complementary to a sequence near the 5′-end of the snRNA called U1. Five U-rich snRNAs (U1, U2, U4, U5, and U6), ranging in length from 107 to 210 nucleotides, participate in RNA splicing. snRNAs are associated with six to ten proteins in small nuclear ribonucleoprotein particles (snRNPs) in the nucleus of eukaryotic cells.

Base-pairing between U2 snRNA and the branch-point sequence occurs in pre-mRNA and this is critical to splicing.

The branch-point A itself, which is not base-paired to U2 snRNA, "bulges out," allowing its 2′-hydroxyl to participate in the first transesterification reaction of RNA splicing. Assembly of a spliceosome begins with the base-pairing of U1 and U2 snRNAs, to the pre-mRNA (Figure 1.14). Extensive base-pairing between the U4, U5 and U6 snRNPs results in U4/U6/U5 complex. This complex then associates with the previously formed complex consisting of a pre-mRNA base-paired to U1 and U2 snRNPs to yield a spliceosome.

The rearranged spliceosome then catalyses the two transesterification reactions that result in RNA splicing. After the second transesterification reaction, the ligated exons are released from the spliceosome while the lariat intron remains associated with the snRNPs. This final intron–snRNP complex is unstable and hence dissociates. The individual snRNPs released participate in a new cycle of splicing.

Polyadenylation

Pre-mRNAs are cleaved at specific 3′ sites and are rapidly polyadenylated. All mRNAs except histone mRNAs have a 3′ poly(A) tail. At least 6 proteins are identified to take part in polyadenylation. Cleavage and polyadenylation specificity factor (CPSF)(160, 100, 73 and 30 kDa protein complex) binds AAUAAA sequence and is involved in both endonucleolytic cleavage and poly(A) addition (Figure 1.15). The 160- and 30-kDa proteins appear to be involved in sequence recognition.

Cleavage stimulation factor (CStF) (77, 64 and 50 kDa protein complex) binds the downstream elements and is required only for the cleavage event. The 64-kDa protein appears to have the sequence recognition and ribonucleoprotein (RNP) binding domain. CStF and CPSF interact and stabilize each other at the site of cleavage. (Interaction between CStF and the GU-rich downstream poly(A) signal stabilizes the multiprotein complex.)

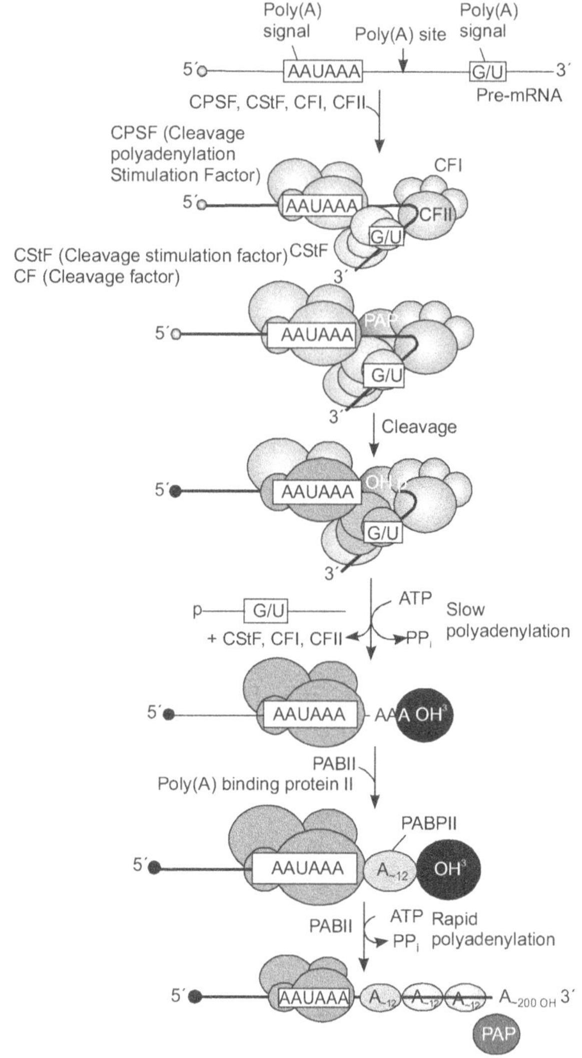

Figure 1.15   Polyadenylation

The role of the cleavage factors (CFI and CFII) has not been defined. Poly(A) polymerase (PAP) is involved in poly(A) tail synthesis using ATP and RNA 3′-OH (after cleavage).

Poly(A) binding protein II (PAB II) binds the growing poly(A) tail and accelerates its extension by poly(A) polymerase.

Addition of the first 12 A residues occurs slowly followed by a rapid addition of up to 200–250 A residues. Many copies of poly(A) binding protein help in this process.

Functions of polyadenylation

1. **mRNA stability** Polyadenylation is thought to contribute to the potential for stability of mRNA by protecting the 3'-end (PAB bound to As) from 3'-exonucleolytic degradation.

2. **Translation** Short-tailed RNAs are poor substrates for translation whereas RNAs with longer tails are better recognized for translation.

TRANSLATION

Translation or protein synthesis comprises three steps:

1. Initiation
2. Elongation
3. Termination

Initiation

During initiation, all the components necessary for protein synthesis assemble. The following are the components

1. mRNA
2. the small and large subunits of the ribosome
3. a set of three proteins called initiation factors
4. initiator tRNA with *N*-formylmethionine attached (fMet-tRNA-fMet) and (5) guanosine triphosphate (GTP).

Initiation comprises three major steps. First, mRNA binds to the small subunit of the ribosome. The rRNA sequence present in the small subunit of ribosome (conserved) has complementarity with the Shine-Dalgarno sequence in prokaryotes and UTR (untranslated region) of eukaryotes and helps in ribosomal binding. Second, initiator tRNA binds to the mRNA through base-pairing between the codon and anticodon. The binding of an amino acid to a tRNA requires the presence of a specific aminoacyl-tRNA synthetase and ATP. The amino acid is attached by its carboxyl end to the 3′-end of the tRNA. Third, the large ribosome joins the initiation complex.

Large subunit of ribosome has three sites that can be occupied by tRNAs—the aminoacyl or A site, the peptidyl or P site, and the exit or E site (Figure 1.16). The initiator tRNA (fMet-tRNA-fMet) immediately occupies the P site leaving the A site for the second tRNA.

Elongation

The next stage in protein synthesis is elongation, in which amino acids are joined to create a polypeptide chain. Elongation requires

1. the 70S complex,
2. tRNAs charged with their amino acids,
3. several elongation factors (EF-Ts, EF-Tu, and EF-G) and
4. GTP.

Elongation occurs in three steps. The first step (Figure 1.16b) is the delivery of a charged tRNA (tRNA with its amino acid attached) to the A site. This requires the presence of elongation factor Tu (EF-Tu), elongation factor Ts (EF-Ts), and GTP. EF-Tu first joins with GTP and then binds to a charged tRNA to form a three-part complex. This three-part complex

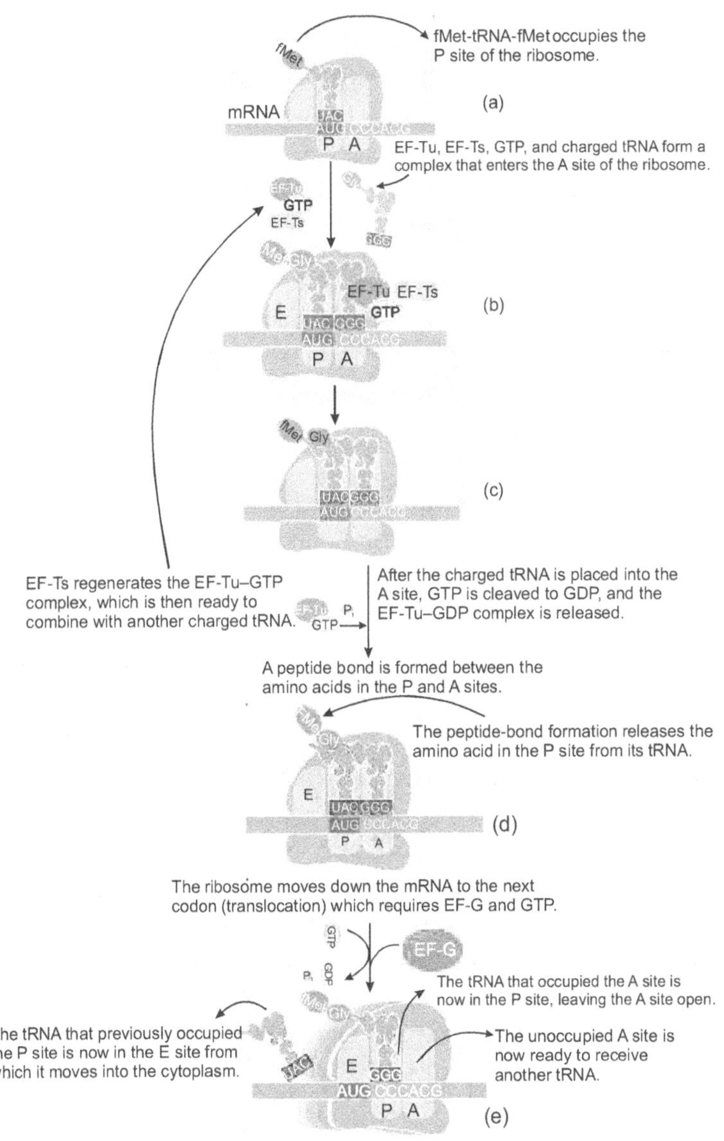

Figure 1.16 The elongation step of translation

enters the A site of the ribosome, where the anticodon on the tRNA pairs with the codon on the mRNA. After the charged tRNA is positioned in the A site, GTP is cleaved to GDP, and the EF-Tu–GDP complex is released (Figure 1.16c). Factor EF-Ts converts EF-Tu–GDP to EF-Tu–GTP. In eukaryotic cells, a similar set of reactions delivers the charged tRNA to the A site.

The second step of elongation is the creation of a peptide bond between the amino acids that are attached to tRNAs in the P and A sites (Figure 1.16d). The formation of this peptide bond releases the amino acid in the P site from its tRNA. The activity responsible for peptide bond formation in the ribosome is referred to as peptidyltransferase. For many years, the assumption was that this activity is carried out by one of the proteins in the large subunit of the ribosome. Evidences, however, now indicate that the catalytic activity is a property of the rRNA in the large subunit of the ribosome, which acts as a ribozyme.

The third step in elongation is translocation, (Figure 1.16e), the movement of the ribosome down the mRNA in the $5' \rightarrow 3'$ direction. This step positions the ribosome over the next codon and requires the elongation factor G (EF-G) and the hydrolysis of GTP to GDP. Then the ribosome shifts so that the tRNA that previously occupied the P site now occupies the E site, from which it moves into the cytoplasm where it may be recharged with another amino acid. Translocation also causes the tRNA that occupied the A site (which is attached to the growing polypeptide chain) to be in the P site, leaving the A site open. Thus, the progress of each tRNA through the ribosome during elongation can be summarized as follows:

Cytoplasm $\rightarrow$ A site $\rightarrow$ P site $\rightarrow$ E site $\rightarrow$ Cytoplasm

As discussed earlier, the initiator tRNA is an exception. It attaches directly to the P site and never occupies the A site.

After translocation, the A site of the ribosome is empty and ready to receive the tRNA specified by the next codon. The ribosome moves down the mRNA in the $5' \rightarrow 3'$ direction, adding amino acids one at a time according to the order specified by the mRNA's codon sequence.

Termination

Protein synthesis terminates when the ribosome translocates to a termination codon. Because there are no tRNAs with anticodons complementary to the termination codons, no tRNA enters the A site of the ribosome when a termination codon is encountered (Figure 1.17).

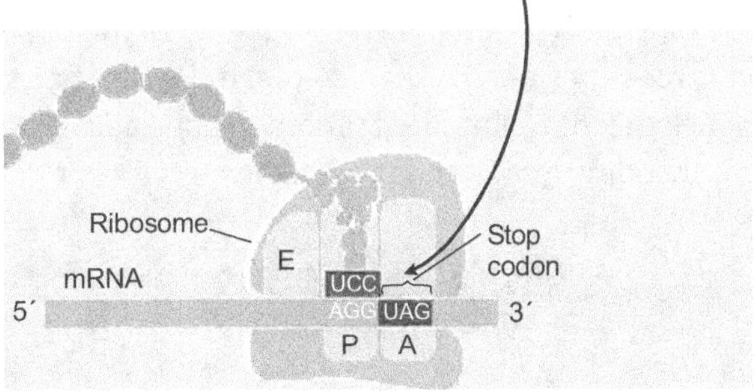

Figure 1.17   Termination process in protein synthesis

When the A site of ribosomal unit reads the termination codon, it recruits the releasing factor (RF) which does not have any amino acid. The delivery of RF to A site stoops the translation process by releasing both small and large subunits of ribosomes from mRNA. Thus protein synthesis is terminated.

The Genetic Code

One amino acid is encoded by three consecutive nucleotides in mRNA, and each nucleotide can have one of four possible bases (A, G, C, and U) at each nucleotide position thus permitting 43 to 64 possible codons. Three of these codons are stop codons, specifying the end of translation. Thus, 61 codons called sense codons code for amino acids. Because there are 61 sense codons and only 20 different amino acids commonly found in proteins, the code contains more information than is needed to specify the amino acids and is said to be a degenerate code.

Codon Usage

The term "codon usage" describes the selective and non-random use of synonymous codons in protein synthesis. There are species-specific patterns of codon usage. In higher plants, monocotyledonary (monocot) and dicotyledonary (dicot) species differ significantly from each other. Monocots show a distinct preference for XXC/G codons and have a higher frequency of XCG but a lower frequency of XTA than dicots. It has also been noted that the genes encoding abundant proteins show more codon bias than those encoding less abundant proteins.

REVIEW QUESTIONS

1. Describe the organization of DNA into chromosomes.

2. Give an outline of the semiconservative mode of DNA replication.

3. Describe the genome organization in plants.

4. Give an account of the processing of mRNA.

5. Explain protein synthesis at the molecular level.

6. Define genetic code.

2

CHLOROPLAST AND ITS GENOME

INTRODUCTION

Chloroplasts are the organelles responsible for photosynthesis. Plants are the only photosynthetic organisms to have leaves (not all plants have leaves). The leaf is the primary photosynthetic organ of typical land plants. The leaf has an extensive surface area exposed to the sun in order to absorb the maximum amount of light (there are, of course, adaptive differences among plants). The major portion of the leaf is the mesophyll, which actually contains the photosynthetic cells, which are the parenchyma cells and each of these parenchyma cells has lots of chloroplasts. The raw materials of photosynthesis, water and carbon dioxide, enter the cells of the leaf, and the products of photosynthesis, sugar and oxygen, leave the leaf. The site of photosynthesis is the chloroplast in leaves. In addition, chloroplasts synthesize amino acids, fatty acids, and the lipid components of their own membranes. The reduction of nitrite (NO_2^-) to ammonia (NH_3), an essential step in the incorporation of nitrogen into organic compounds, also occurs in chloroplasts.

Chloroplasts are in many respects similar to mitochondria. Both chloroplasts and mitochondria function to generate metabolic energy. Evolved by endosymbiosis, they contain their own genetic systems and replicate by division. However, chloroplasts are larger and more complex than mitochondria, and they perform several critical tasks in addition to the generation of ATP.

CHEMISTRY OF CHLOROPLASTS

Chloroplasts are rich in proteins, lipids, carbohydrates, nucleic acids, pigments and minerals. Mature chloroplasts contain 35 per cent of total proteins, of which quinines, flavins, lipoproteins and tocopherol form a major portion. Chloroplasts synthesize starch as the main product and store carbohydrates in plenty. DNA and RNA are associated with chloroplasts. Haeme and chlorophylls are major pigments in the chloroplasts, but "chlorophyll *a*" is predominant. A number of elements in free or bound forms are also found in these organelles, which include copper, iron, magnesium and manganese. Magnesium is an important component of chloroplast.

STRUCTURE

The chloroplasts (Figure 2.1) are large organelles (5 to 10 μm long) that, like mitochondria, are bounded by a double membrane called the chloroplast envelope. In addition to the inner and outer membranes of the envelope, chloroplasts have a third internal membrane system, called the thylakoid membrane. The thylakoid membrane forms a network of flattened discs called thylakoids which are frequently arranged in stacks called grana (granum). The interior of the thylakoids is called the lumen and the exterior (volume bounded by the inner membrane) is called the stroma. The grana are in liquid-filled compartments called the stroma. Chloroplast molecules embedded in the thylakoid membranes give the leaf its green colour.

Because of this three-membrane structure, the internal organization of chloroplasts is more complex than that of mitochondria. In particular, their three membranes divide chloroplasts into three distinct internal compartments.

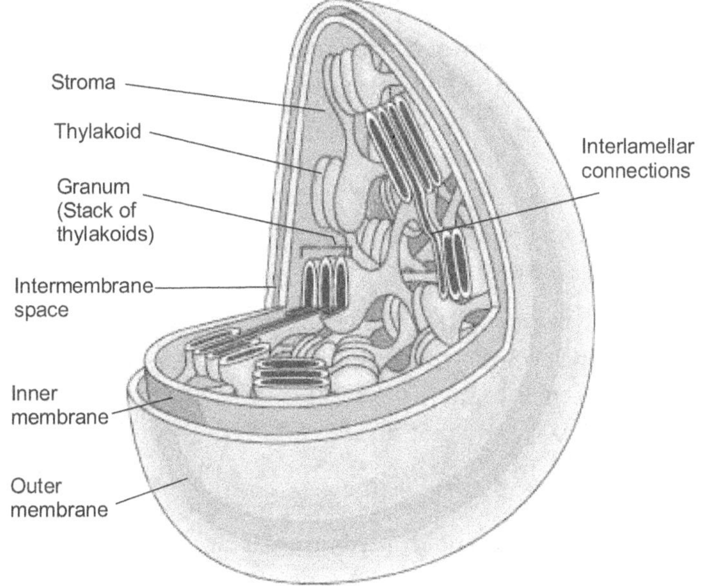

Figure 2.1 Structure of a chloroplast

1. the intermembrane space between the two membranes of the chloroplast envelope

2. the stroma, which lies inside the envelope but outside the thylakoid membrane

3. the thylakoid lumen

Despite this greater complexity, the membranes of chloroplasts have clear functional similarities with those of mitochondria, one of the roles being the generation of ATP. The outer membrane of the chloroplast envelope, like that of mitochondria, contains porins and is therefore freely permeable to small molecules. In contrast, the inner membrane is impermeable to ions and metabolites, which are therefore able to enter chloroplasts only via specific membrane transporters. These properties of the inner and outer membranes of the chloroplast envelope are similar to the inner and outer membranes of mitochondria. In both the cases, the inner membrane restricts the passage of molecules between the cytosol and the interior of the organelle. The chloroplast stroma is also equivalent in function to the mitochondrial matrix. It contains the chloroplast genetic system and a variety of metabolic enzymes, including those responsible for the critical conversion of CO_2 to carbohydrates during photosynthesis.

The major difference between chloroplasts and mitochondria, in terms of both structure and function, is the thylakoid membrane. This membrane is of central importance in chloroplasts, where it fills the role of the inner mitochondrial membrane in electron transport and the chemiosmotic generation of ATP. The inner membrane of the chloroplast envelope (which is not folded into cristae) does not function in photosynthesis. Instead, the electron transport system of chloroplast is located in the thylakoid membrane, and the protons are pumped across this membrane from the stroma to the thylakoid lumen. The resulting electrochemical gradient then drives ATP synthesis as protons cross back into the stroma. In terms of its role in generation of metabolic energy, the thylakoid membrane of chloroplasts is thus equivalent to the inner membrane of mitochondria.

PIGMENTS

Photosynthetic pigments are embedded in the thylakoid membranes. The pigment is a substance that absorbs light. Pigment molecules have electrons in relatively low-energy orbitals. The colour of the pigment comes from the wavelengths of light reflected (in other words, those not absorbed). The various pigments include chlorophyll *a*, chlorophyll *b*, carotenoids and xanthophylls.

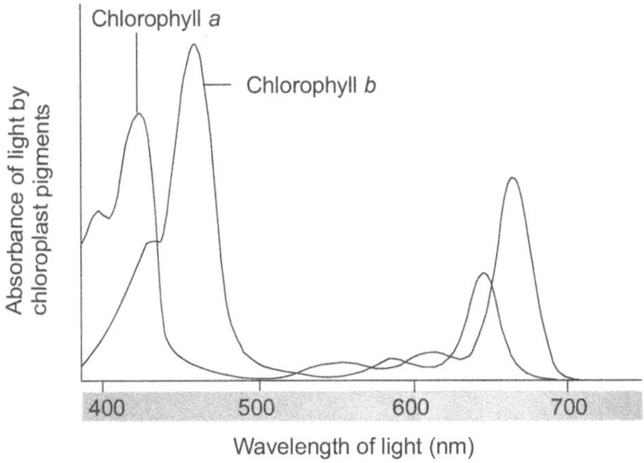

Figure 2.2 Absorption of wavelength of light by chloroplast pigments

Chlorophyll *a* is the primary photo pigment. Chlorophyll *a* absorbs its energy more from the violet-blue (400–500 nm) and reddish orange-red wavelengths (600–700 nm), and little from the intermediate (green-yellow-orange) (Figure 2.2). It must be present to carry out photosynthesis. All photosynthetic organisms (plants, certain protistans, prochlorobacteria, and cyanobacteria) have chlorophyll *a*. The electrons in these pigments absorb light energy and because of this they get boosted from their ground state to their excited state. When excited, they actually get released from the molecule. This initiates the photosynthesis process. Hence, it is called as primary photo pigment.

Accessory pigments include chlorophyll *b* (also *c*, *d*, and *e* in algae and protistans), xanthophylls and carotenoids. They absorb energy that chlorophyll *a* does not absorb (Figure 2.2), passing it along to chlorophyll *a* and thus are supplementary light receptors. Carotenoids may be yellow, red or purple and the most important are β-carotene, which is red-orange and lutein which is yellow.

The different types of plastids are frequently classified according to the kinds of pigments they contain. Chloroplasts are so named because they contain chlorophyll. Chromoplasts lack chlorophyll but contain carotenoids; they are responsible for the yellow, orange, and red colours of some flowers and fruits, although their precise function in cell metabolism is not clear. Leucoplasts are non-pigmented plastids, which store a variety of energy sources in non-photosynthetic tissues. Amyloplasts and elaioplasts are examples of leucoplasts that store starch and lipids, respectively.

FUNCTION OF CHLOROPLAST—PHOTOSYNTHESIS

Photosynthesis is the chief function of chloroplasts. Light is central to the life of the plant. Plant cells convert light into chemical signals that affect a plant's life cycle. Photosynthesis is the process by which chloroplast-bearing organisms transform solar light energy into chemical bond energy. It is the most important chemical process on earth. It provides food and oxygen for virtually all organisms. Autotrophs (plants, algae and some bacteria) make their own food and most use photosynthesis to do it. All life depends on autotrophs directly or indirectly for food.

Water enters the root and is transported up to the leaves through specialized plant cells known as xylem vessels. Land plants must guard against drying out and so have evolved specialized structures known as stomata to allow gas to enter and leave the leaf. Carbon dioxide cannot pass through the protective waxy layer covering the leaf (cuticle), but it can enter the leaf through the stoma (the singular of stomata), flanked by two guard cells. Likewise, oxygen produced during photosynthesis can only pass out of the leaf through the opened stomata. Unfortunately for the plant, while these gases are moving between the inside and outside of the leaf, a great deal of water is also lost. Cottonwood trees, for example, will lose 100 gallons (about 450 dm^3) of water per hour during hot days.

Photosynthesis is the reduction of carbon dioxide into carbohydrate via the oxidation of energy carriers (ATP and NADPH) and the reaction can be summarized as

$$6CO_2 + 6H_2O \xrightarrow[\text{Chlorophyll}]{\text{Light}} C_6H_{12}O_6 + 6O_2$$

The total photosynthetic process is made up of two series of reactions. These two processes are chemically and physically separate (Table 2.1). Two stages of photosynthesis are:

i. light-dependent reactions or photosynthetic reactions (also called the light reactions or photo reactions) capture the energy of light and use it to make high-energy molecules.

ii. light-independent reactions, also called the Calvin–Benson cycle, and formerly known as the dark reaction. However, dark reactions are now called **carbon reactions,** as they are also regulated by light.

The two types of reactions that occur in the chloroplast are diagrammatically represented in Figure 2.3.

Light Reactions

These reactions occur in four steps in the thylakoid membrane.

1. Primary photo-event—A photon of light is captured by a pigment.

2. Charge separation—Excitation energy is transferred to the reaction centre, and an energetic electron is transferred to an acceptor molecule.

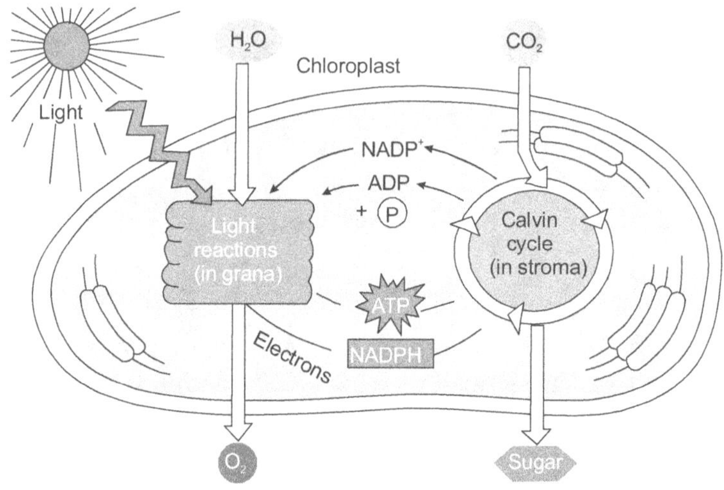

Figure 2.3   Photosynthetic processes in chloroplast

Table 2.1   Light reactions and dark reactions

Light-dependent reactions	Light-independent reactions or Calvin–Benson cycle or dark reactions
Take place in thylakoid membranes using chlorophyll and other pigments	Take place in stroma of chloroplast
Capture light energy and release O_2	Fix CO_2
Convert light energy to chemical energy via electron transport	Use energy from light-dependent reactions to synthesize carbohydrates
Store energy in carriers : ATP and NADPH	Use ATP and NADPH
Need water	Need water

3. Electron transport—Proton pumping and reduction of $NADP^+$.

4. Chemiosmosis—Photophosphorylation and ATP production.

In other words, the light strikes chlorophyll *a* in such a way as to excite electron to a higher energy state. In a series of reactions the energy is converted (along an electron transport process) into ATP and NADPH. Water is split in the process, releasing oxygen as a by-product of the reaction. The reactions are given by

$$2H_2O + 2NADP^+ \rightarrow 2NADPH + 2H + O_2$$
$$ADP + Pi \rightarrow ATP + H_2O$$

The ATP and NADPH produced in the light reactions are used to make C-C bonds in the light-independent process (Calvin cycle). Almost all the chemical processes that make up the light reactions of photosynthesis are carried out by four major protein complexes embedded on the thylakoid membrane.

1. photosystem II

2. cytochrome b_6f complex

3. photosystem I

4. ATP synthase

These four integral membrane complexes are vectorially oriented in thylakoid membrane. Therefore the electron transport reactions take place in the thylakoid membrane. To enable the reduction of $NADP^+$, the light energy must be captured to elevate the redox potential of the electrons of water. The elevation of the redox potential of these electrons is the role of photosynthetic electron transport.

The flow of electrons down the ETC, when represented based on the redox potential of the complexes, resembles the letter Z and is therefore known as the Z scheme. Figure 2.4 shows the Z scheme, but a detailed description of all the complexes involved in the ETC are beyond the scope of this book.

Events of the light reaction   In the primary photoevent, a photon of light (*hv*) falls on PS II and is absorbed, thereby increasing its redox potential. PS II reduces plastoquinone, an electron carrier in the thylakoid membrane, and gets oxidized causing charge separation to occur and setting into motion the electron transport process. Oxidized PS II has a very low redox potential. P680 (PS II), now short of an electron, actively splits water molecules. This splitting in the presence of light is called **photolysis**. The missing electron is procured via a protein, which contains 4 Mn atoms (water splitting complex). The role of the Mn is to pull electrons away from water. The Mn–protein complex splits the two water molecules and takes up four electrons, thereby releasing oxygen. The H^+ released from water is liberated on the inner side of the thylakoid membrane.

The reduced plastoquinone reduces the cytochrome b_6f complex. The H^+ ions that are pumped into the thylakoid lumen cytochrome b_6f complex reduce PS I (P700). PS I absorbs a photon of light. Therefore the redox potential becomes very high. With high redox potential, ferredoxin can reduce $NADP^+$ to NADPH.

While a fraction of captured light is used to reduce $NADP^+$ to NADPH, the other fraction of light is used for light-dependent ATP synthesis, which is known as photophosphorylation. A proton gradient is formed across the thylakoid membrane, by pumping protons into the thylakoid lumen, during the electron transport process as described earlier. These protons now flow back out through the CF_1-CF_0 ATP synthase down the gradient causing the synthesis of ATP from ADP and Pi. This is also known as non-cyclic photophosphorylation because, electrons expelled from PS II is used up in NADP reduction process and not returned back. It occurs when both NADPH and ATP are required by the plant.

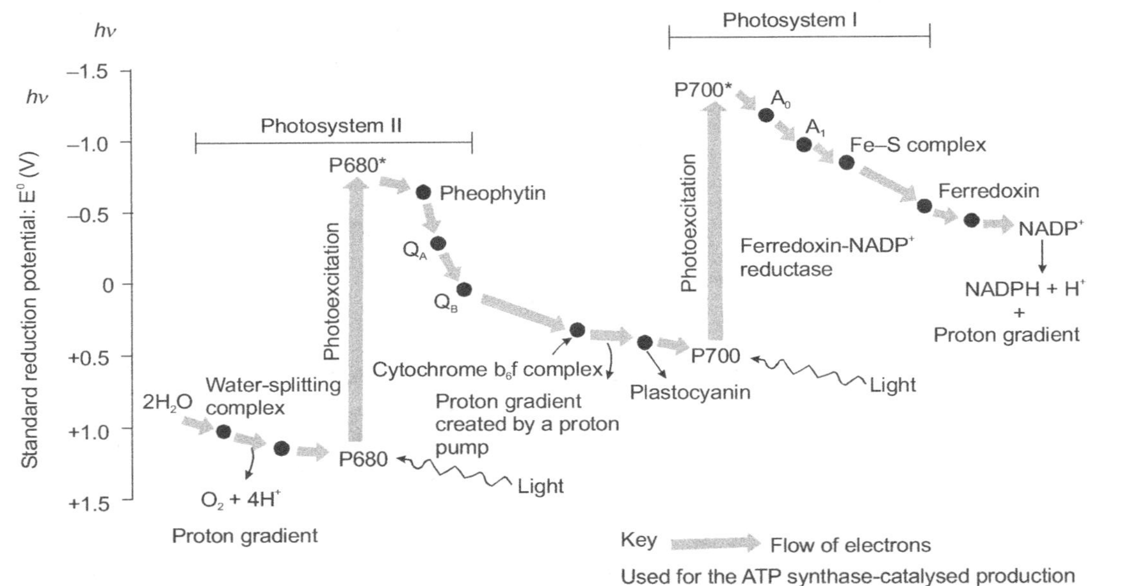

Figure 2.4   Flow of electrons based on the redox potential of the complexes (Z scheme) P700 or PS I, P680 or PS II

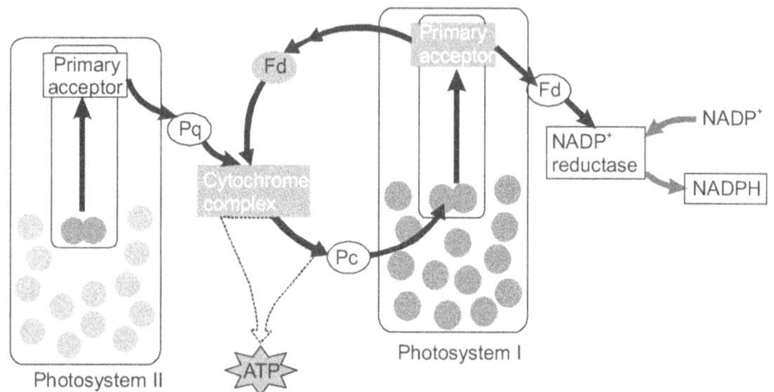

Figure 2.5   Cyclic phosphorylation showing cyclic flow of electrons

Sometimes electrons do not go through the whole chain of electron acceptors in photosystem I, but being recycled through the cytochrome complex that precedes photosystem I. When this occurs (cyclic electron flow), it is termed as cyclic photophosphorylation (Figure 2.5). It happens only in photosystem I. It still contributes to the production of ATP, but not to the production of NADPH.

Thus, in summary the light reactions use energy from the sun to make ATP and NADPH, using a proton pump powered by energy from an electron transport system. The energy-rich molecules ATP and NADPH synthesized in light reaction is used to fix CO_2 in the form of sugars.

Calvin Cycle (Reductive Pentose–Phosphate Pathway)

Carbon-fixing reactions are also known as the dark reactions or light-independent reactions. The most common pattern of the dark reactions is the Calvin cycle. Carbon dioxide enters the single-celled and aquatic autotrophs through non-specialized structures, diffusing into the cells. As mentioned earlier, land plants must guard against drying out (desiccation) and so have stomata to allow the gas to enter and leave the leaf. The Calvin cycle occurs in the stroma of chloroplasts and on the thylakoid surfaces facing the stroma.

It is essential that these reactions follow the light reactions because they require NADPH and ATP. Dark reactions take place in the dark; but they can also take place in the presence of light. They convert the chemical bond energy of NADPH and ATP into the chemical bond energy of glucose. The Calvin cycle requires different levels of ATP and NADPH. It requires 9 ATP for every 6 NADPH. Six molecules of carbon dioxide enter the Calvin cycle, eventually producing one molecule of glucose. The reactions in this process were worked out by Melvin Calvin.

The Calvin cycle is just a cycle in which the same reactions are repeated many times. The starting material should be present and be continuously regenerated in order for this to occur. The starting material is a five carbon (5C) sugar.

The Calvin cycle is divided into three phases (Figure 2.6).

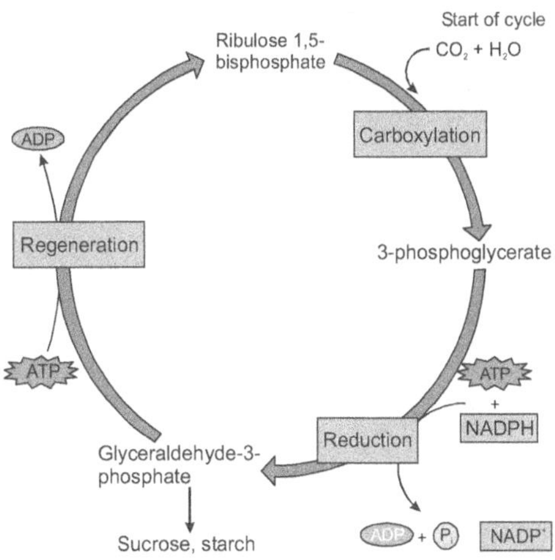

Figure 2.6   Calvin cycle showing the three phases

i. Carboxylation or carbon fixation

ii. Reduction

iii. Regeneration of the starting product

Carboxylation or carbon fixation Carbon dioxide molecules enter the leaf mesophyll through the stoma and go into the parenchyma cells. The CO_2 molecules bind to the 5-C sugar, RuBP (ribulose 1,5-bisphosphate), aided by the enzyme RuBisCO (Ribulose bisphosphate carboxylase/oxygenase). This makes a highly unstable six-carbon sugar, which immediately splits into two three-carbon molecules. This is the key initial step in C3 photosynthesis. The reaction is given by

CO_2 + Ribulose 1,5-bisphosphate (RuBP) + H_2O → 2,3-Phosphoglycerate (PGA)

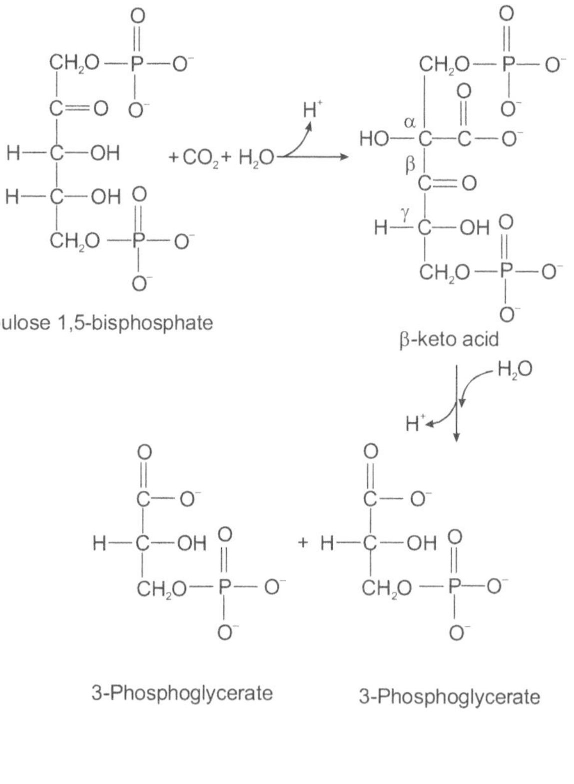

Ribulose 1,5-bisphosphate

β-keto acid

3-Phosphoglycerate

3-Phosphoglycerate

This binding of the five-carbon sugar ribulose 1, 5-bisphosphate (RuBP) molecule and 1C CO_2, yields two three-carbon molecules and therefore known as C3 photosynthesis—the sole type of carbon fixation reaction in most plants.

$$5C + 1C \rightarrow 2 \times 3C \text{ molecules}$$

Reduction   The three-carbon molecules thus generated are phosphorylated by ATP to form 1,3-bisphosphoglycerate (1,3-BPG).

$$\overset{1}{C}OO^- \quad \overset{2}{C}HOH \quad \overset{3}{C}H_2OPO_3^{2-} \;+\; ATP \;\xrightleftharpoons{\text{3-Phosphoglycerate kinase}}\; \overset{1}{C}OPO_3^{2-} \quad \overset{2}{C}HOH \quad \overset{3}{C}H_2OPO_3^{2-} \;+\; ADP$$

BPG receives electrons from NADPH and gets reduced to two glyceraldehyde 3-phosphate (G3P) molecules (3-C each).

$$\underset{CH_2OPO_3^{2-}}{\overset{C-OPO_3^{2-}}{CHOH}} \;+\; NADPH + H^+ \;\xrightleftharpoons{\substack{\text{Glyceraldehyde} \\ \text{3-phosphate} \\ \text{dehydrogenase}}}\; \underset{CH_2OPO_3^{2-}}{\overset{\overset{H}{C}\!\!=\!\!O}{CHOH}} \;+\; NADP^+ + Pi$$

The two triose phosphates, glyceraldehyde 3-phosphate and dihydroxyacetone phosphate (DHAP), are in equilibrium with each other catalysed by the action of triose phosphate isomerase similar to gluconeogenesis.

$$\text{2-Glyceraldehyde 3-phosphate} \;\xrightleftharpoons{\substack{\text{Triosephosphate} \\ \text{isomerase}}}\; \text{Dihydroxyacetone phosphate}$$

The triose phosphates are either converted to starch, which can be stored for later use, or converted to sucrose (Figure 2.6) that can be used immediately.

Regeneration of the starting product Regeneration of RuBP is important for continued carbohydrate synthesis. This is achieved through a series of reactions given below. G3P molecules react to regenerate ribulose 1,5 bisphosphate (the starting compound) and hexose phosphate molecules (precursors of sugar and starch molecules). The process of regeneration involves carbon shuffling reactions given below.

1. Conversion of hexoses and trioses to a four- and a five-carbon sugar is catalysed by the enzyme transketolase.

Fructose 6-Pi + Glyceraldehyde 3-Pi → Erythrose 4-Pi +Xylulose 5-Pi

$$6C + 3C \rightarrow 4C + 5C$$

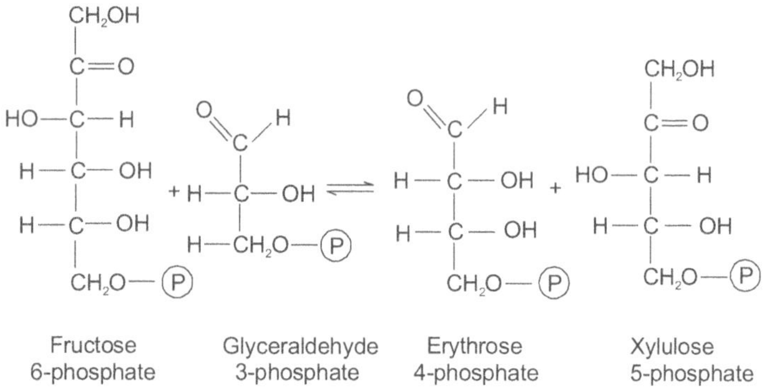

Fructose 6-phosphate	Glyceraldehyde 3-phosphate	Erythrose 4-phosphate	Xylulose 5-phosphate

2. Then aldolase condenses erythrose 4-phosphate with dihydroxyacetone phosphate to form sedoheptulose 1,7-bisphosphate. Sedoheptulose 1,7-bisphosphatase removes the phosphate group from C1 position to form sedoheptulose 7-phosphate.

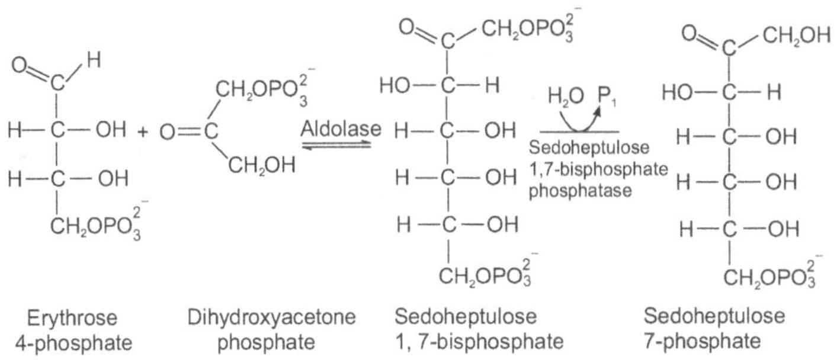

Erythrose 4-phosphate + Dihydroxyacetone phosphate $\xrightarrow{\text{Aldolase}}$ Sedoheptulose 1,7-bisphosphate $\xrightarrow[\text{1,7-bisphosphate phosphatase}]{\text{Sedoheptulose}}$ Sedoheptulose 7-phosphate

$$\text{DHAP} + \text{Erythrose 4-Pi} \rightarrow \text{Sedoheptulose 1,7-bisPi}$$

$$3C + 4C \rightarrow 0C + 7C$$

Transketolase   Sedoheptulose 7-phosphate then reacts with a second molecule of glyceraldehyde 3-phosphate in the presence of transketolase to form ribose 5-phosphate and xylulose 5-phosphate.

Sedoheptulose 7-phosphate + Glyceraldehyde 3-phosphate $\xrightarrow{\text{Transketolase}}$ Ribose 5-phosphate + Xylulose 5′-phosphate

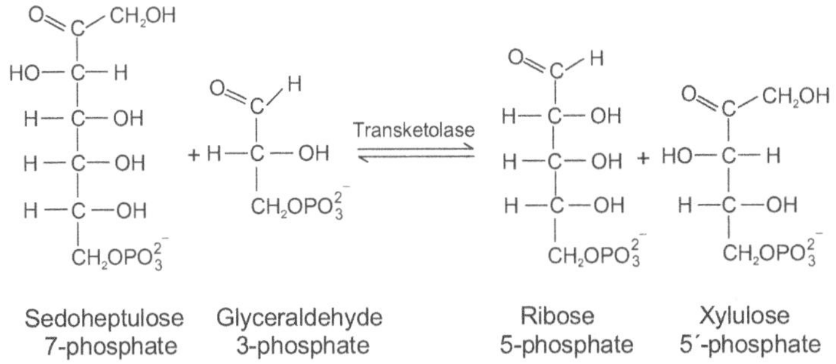

Finally, ribose 5-phosphate is converted into ribulose 5-phosphate by phosphopentose isomerase and xylulose 5-phosphate is converted into ribulose 5-phosphate by phosphopentose epimerase. The last step is the phosphorylation of ribulose 5-phosphate to form ribulose 1,5-bisphosphate. With this, the cycle gets completed.

Photorespiration   On hot summer days, the stomata close to prevent loss of water. This causes a decrease in CO_2

concentration and an increase in O_2 (from light reactions). Thus the Calvin cycle is diverted and photorespiration occurs. RuBisCO, the first enzyme in the Calvin cycle can also accept oxygen as a substrate or in other words also acts as an oxygenase and forms phosphoglycolate (2C) and 3-phosphoglycerate (PGA) instead of two molecules of PGA, causing loss of organic carbon and waste of energy in recycling phosphoglycolate.

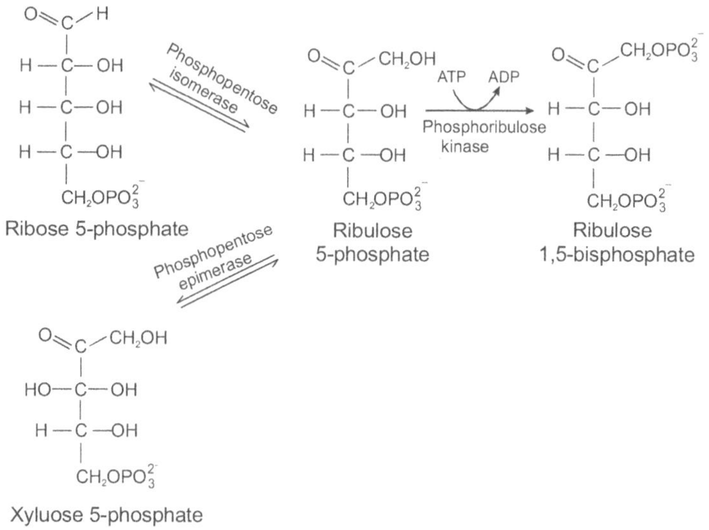

Ribose 5-phosphate

Phosphopentose isomerase

Phosphopentose epimerase

Xyluose 5-phosphate

Ribulose 5-phosphate

ATP   ADP

Phosphoribulose kinase

Ribulose 1,5-bisphosphate

There are a couple of ways that some plants have dealt with photorespiration: C4 metabolism and CAM metabolism.

C4 metabolism C4 plants make 4C molecule (oxaloacetate) instead of 3C phosphoglycerate. CO_2 is fixed in the mesophyll cells into a 4C compound. The 4C compound diffuses into bundle sheath cells where it is cleaved into CO_2 and 3C pyruvate (Figure 2.7); pyruvate returns to the mesophyll and CO_2 flows into the Calvin cycle finally leading to formation of glucose. C4 photosynthesis occurs in most grasses (corn and sugar cane).

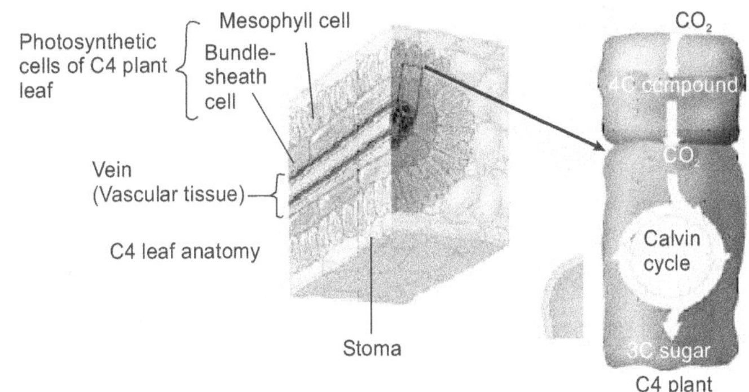

Figure 2.7 C4 metabolism involving four-carbon compounds

CAM (Crussalacean acid metabolism) They open their stomata only at night, i.e., at cool temperature, and water loss is less likely. When their stomata are open, they take in carbon dioxide; convert it into organic acids; and hold the carbon in the form of organic acids, until daylight. When the sun rises, the plant can proceed with the light reactions, and then use the stored carbon in the form of organic acids for the Calvin cycle.

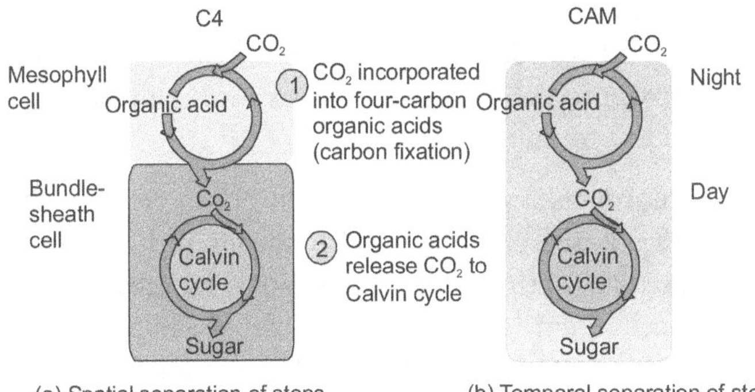

Figure 2.8 Spatial and temporal separation of C4 plants and CAM plants

Thus the C4 plants spatially separate carbon dioxide fixation from Calvin cycle whereas the CAM plants have a temporal (time) separation (Figure 2.8) of these events.

PLASTOME OR CHLOROPLAST DNA (cpDNA)

The presence of DNA in chloroplasts was first demonstrated by Ris and Plaut in 1962. They observed fibrils in electron micrographs of chloroplasts, which could be removed by treatment with deoxyribonuclease. The photosynthetic cells of higher plants contain on an average 50 chloroplasts, each with 10–20 copies of the plastome. The chloroplast genes are repeated approximately 500–1000 times. More recently, multiple copies of the plastid genome have been visualized by light microscopy, using DNA-binding fluorescent dyes. In very young leaves, chloroplast may contain 200 or more copies of the plastome.

In higher plants, chloroplast DNA (cpDNA) exists as a double-stranded circular molecule. Unlike nuclear DNA, it does not contain 5-methyl cytosine and is not complexed with histones. The GC content varies from 36–40% in different species. Purified chloroplast DNA is estimated to have a molecular weight of 120–150 kb. The DNA is replicated in a semiconservative fashion. Initially the DNA forms a theta structure that spreads outward to form daughter replicons. Later, DNA synthesis switches over to a rolling circle mechanism resulting in multiple copies of the genome.

In chloroplasts of maize and pea, DNA replication begins at two sites about 7 kb apart and proceeds in both directions. Multiple copies of the genome may also be synthesized by rolling circle mechanism (Kolodner and Tewari, 1975). Chloroplast division seems to be related to cell size, and the number of chloroplasts per cell increases in proportion to the cell volume. However, DNA synthesis does not keep pace with the process of plastid division and older chloroplasts may only contain 10–20 copies of the plastome. The DNAs of chloroplast and chromoplast are identical.

The chloroplast genomes from a number of plant and algal species have been sequenced. Many chloroplast genes are organized into operon-like clusters. A typical chloroplast genome of a vascular plant codes for 4 rRNA genes, 30 to 35 tRNA genes, a number of ribosomal proteins and some of the proteins engaged in photosynthetic and non-photosynthetic processes. A key protein encoded by cpDNA is ribulose 1,5-bisphosphate carboxylase/oxygenase which participates in the fixation of carbon in photosynthesis. RuBisCO makes up about 50% of the protein found in green plants and is therefore considered the most abundant protein on earth. It is a complex protein consisting of eight identical small subunits. The large subunit is encoded by chloroplast DNA, and the small subunit by nuclear DNA. The circular chloroplast genome has genes on both its strands. One of the characteristic features of the chloroplast genome is the presence of a large inverted repeat. In rice, this repeat includes genes for 23S rRNA, 4.5S rRNA, and 5S rRNA as well as several genes for tRNA and proteins.

Chloroplast DNA consists of many non-coding sequences. Introns are found in many of the chloroplast genes.

STRUCTURE

Chloroplasts, like bacteria, contain 70S ribosomes. Each 70S ribosome consists of two subunits, a large 50S subunit and a smaller 30S subunit. The small subunit includes a single RNA molecule that is 16S in size. The larger 50S subunit includes three rRNA molecules: a 23S rRNA, a 5S rRNA and a 4.5S rRNA. In bacterial ribosomes, the large subunit possesses only two rRNA molecules, which are 23S and 5S in size. Chloroplast ribosomes contain about 60 distinct ribosomal proteins, encoded by the nucleus and plastome that are distributed between the two subunits.

The rRNA genes for 70S ribosomes were first located on the plastome. More recently, the genetic arrangement has been studied by restriction enzyme mapping and DNA sequencing. The genes for stable rRNAs occur in a cluster, interspersed with tRNA genes. The arrangement resembles that of *E. coli*. The DNA sequences of the 16S rRNA from maize and tobacco consist of 1491 and 1489 nucleotides respectively. They show 96% homology with each other and 75% homology with 16S rRNA from *E. coli*. The 16S, 23S and 4.5S rRNA sequences in chloroplasts, together with the tRNA and introns in the spacer region between the 16S and 23S genes, are transcribed as a polycistronic precursor RNA which is subsequently processed or modified to produce mature rRNAs and tRNAs. Chloroplast precursor rRNA contains one copy each of the 16S, 23S and 4.5S rRNA sequences, but apparently not the 5S sequence.

The chloroplast DNA of tobacco consists of a double-stranded circle of 155,844 base pairs and encodes at least 146 genes (Sugiuna, 1992) (Figure 2.9).

The rRNA gene sequence is flanked by a pair of inverted repeat regions (IR$_A$, IR$_B$). The inverted repeat region is 25,339 bp long and contains 24 genes. Chloroplast genes in tobacco that encode stable RNAs include those for 23S, 16S, 4.5S and 5S rRNA and at least 30 tRNAs. The 28 stromal proteins encoded by the tobacco plastome include 19 different ribosomal proteins such as the subunits of RNA polymerase (*rpo*A, B, C) and the large subunit of RuBisCO (*rbc*L). Nineteen genes for thylakoid proteins have been identified including two (*psa* A, *psa* B) for the P700 (reaction centre apoprotein of PS I) and eight (*psb* A, B, C, D, E, F, G, H) for the polypeptides of PS II. In addition, three genes (*pet* A, *pet* B, *pet* D) have been identified for polypeptides of the cytochrome b$_6$f complex and 6 out of 9 polypeptides (*atp* A, B, E, F, H, I) of the ATP synthase complex. The plastome also has 7 genes for the polypeptides that form the subunits of NADP dehydrogenase.

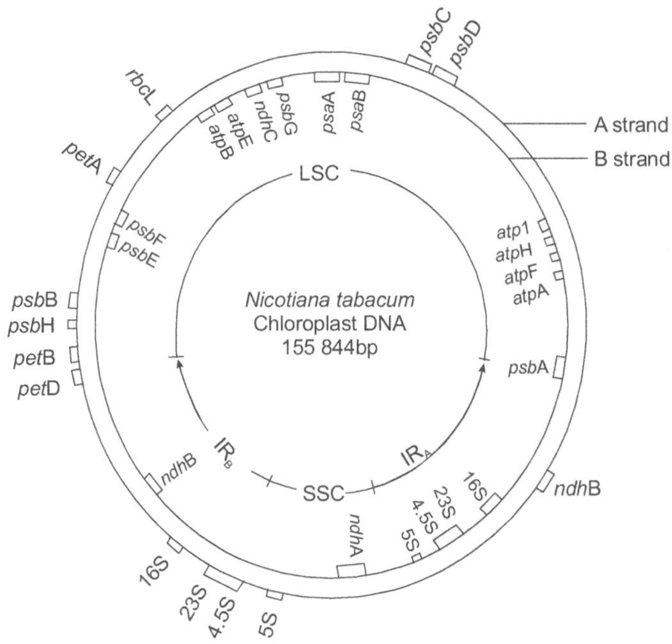

Figure 2.9   Tobacco chloroplast genome (represented with few genes). It contains pair of inverted repeats (IR$_A$ and IR$_B$) which are separated by the small and large single-copy regions (SSC and LSC, respectively). Genes on the outside circle (A strand) are transcribed clockwise, those on the inside (B strand) are transcribed anticlockwise.

GENE EXPRESSION IN CHLOROPLASTS

The transcription and translation of chloroplast genes are similar to that of bacteria.

Transcription

Promoters found in cpDNA possess sequences similar to the –10 and –35 regions of bacterial promoters. Most genes in cpDNA are transcribed in groups. Only a few genes have their own promoters and are transcribed as separate mRNA molecules. The RNA polymerase that transcribes cpDNA is similar to that of bacteria. Like bacterial mRNAs, chloroplast mRNAs are not capped at the 5´-ends and poly (A) tails are not added to the 3´-ends. Many chloroplast mRNAs have a Shine-Dalgarno sequence in the 5´ untranslated region, which serves as a ribosome-binding site.

Transcription termination   Sequences in the untranslated region at the 3´-end of some chloroplast genes like the RuBisCO form stem-loop structures where the mRNA ends, approximately 98 nucleotides downstream from the termination codon for protein synthesis. It is believed that these features have significance for termination of transcription. Termination signals with inverted repeats occurring just before the transcription stop sites, have been found in a number of chloroplast genes.

Translation

Plastids contain tRNA synthetase enzymes and unique tRNAs that are not found in the plant cytosol. Initiation of protein synthesis occurs with formyl methionine. The tRNAs of chloroplasts are capable of being charged with all of the 20 amino acids. Isolated chloroplast can carry out apparently normal protein synthesis. The 30 or so different tRNA genes on the plastome are believed to be sufficient for translating chloroplast-encoded RNA. Chloroplast, unlike mitochondria, uses the normal genetic code. The differences between the ribosomes of chloroplast and cytosol are given in Table 2.2.

Chloroplast protein synthesis has been studied *in vitro* by re-suspending the isolated organelles in a medium without osmoticum, thus disrupting the envelopes partially and making them permeable to externally added ATP and other substrates.

Table 2.2 Differences between the ribosomes of chloroplast and cytosol

Characteristics	Chloroplast				Cytosol			
Initiating amino acid for protein synthesis	Formylmethionine				Methionine (Also formyl methionine)			
Antibiotics which specifically inhibit protein synthesis	D-therochloramphenicol lincomycin				Cycloheximide			
Sedimentation coefficient of ribosomes	70S				80S			
Sedimentation coefficient of ribosome subunits	Large subunit 50S		Small subunit 30S		Large subunit 60S		Small subunit 40S	
Sedimentation coefficient of rRNAs	23S	5S	4.5S	16S	25S	5.8S	5S	18S
Approximate number of nucleotides	2810	122	103	1490	3580	157	120	1926

Intact chloroplasts are capable of taking up added amino acids from the medium and incorporating these into proteins, using ATP generated by photophosphorylation. To date, there is no evidence that mRNAs can enter chloroplasts from the cytosol, so it is assumed that they are all transcribed from chloroplast DNA.

Regulation of Chloroplast Gene Expression

In general, transcription of chloroplast DNA occurs at high rates during the rapid phase of leaf development and then declines as the leaf cells mature. Effect of light on plastid gene expression has been studied in detail. In general, light acting through the phytochrome system stimulates the accumulation of large subunit of RuBisCO (*rbc*L) and small subunit of RuBisCO (*rbc*S) mRNA and protein, although the extent of their transcription and translation varies significantly in different plants. A dicistronic transcript for *psa*A *psa*B is present in barley etioplasts in the dark but appears not to be translated unless illuminated.

psbA gene has been studied extensively. It encodes a 32-kDa quinone-binding protein of PS II. The mRNA for *psbA* is present at low levels in etioplasts and accumulates rapidly following illumination. Although light can modulate rates of transcription of chloroplast genes, other factors such as mRNA stability, rate of mRNA translation and turnover rate of specific proteins are also important in determining the expression of specific genes.

Some nuclear genes encoding plastid functions are listed below:

1. RuBisCO small subunit (*rbc*S)
2. RuBisCO large subunit binding protein
3. Subunits of ATP synthase (*atp*C,D,G)
4. Reiske Fe-S centre of cyt b$_6$f complex (pet C)

5. Polypeptides of the O_2 evolving complex of PS II

6. Ferredoxin (pet F)

7. Ferredoxin NADP oxidoreductase

8. Plastocyanin (pet E)

9. Chlorophyll *a/b* binding (cab) proteins of light-harvesting complexes LHC I and LHC II

10. RNA polymerase

11. Some ribosomal proteins

12. tRNA synthetases

13. Some genes for CO_2 metabolism, starch synthesis, chlorophyll synthesis, and carotenoid synthesis.

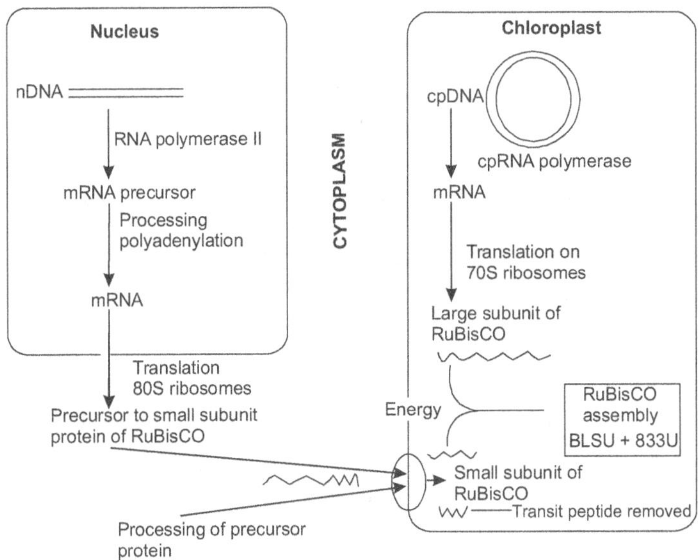

Figure 2.10 A model of the synthesis, transport and assembly of ribulose bisphosphate carboxylase/oxygenase (RuBisCO)

Among the nuclear genes that have been cloned and characterized are those for the small subunit of RuBisCO (*rbcS*), and other proteins are listed above. It indicates that a mechanism exists for transporting macromolecules from the nucleus or the cytosol into chloroplasts (Figure 2.10).

All nuclear encoded genes for chloroplast proteins have mRNA with a 5′ cap and a 3′ poly(A) sequence. The signal sequences attached to the N-terminus of the mature proteins contain information for recognition and transport of the proteins by the chloroplast double envelope membranes.

The precursor protein of RuBisCO small subunit is recognized by a receptor on the chloroplast envelope and crosses the double membrane by a post-translational mechanism, which requires ATP. Once inside the envelope, the transit peptide is processed by a soluble protease in the stroma.

The large subunit of RuBisCO binds in a protein (called "large subunit binding protein" [BP]) that binds Mg^{2+} and ATP. It has been proposed that BP is required for the assembly of newly synthesized RuBisCO subunits into the holoenzyme.

One of the processing enzymes that first encounters the transit sequence after it has bound to the receptor on the chloroplast envelope is a metalloprotease in the stroma that has a molecular weight of 180 kDa.

The location of plastocyanin is the inner face of the lumen of the thylakoids. There are two functional domains in the pre-sequence. Mature plastocyanin consists of 99 amino acids and the pre-sequence contains 69 amino acids. After entry of pre-plastocyanin into chloroplasts, the stromal protease removes approximately two-thirds of the pre-sequence in a two-stage process. The remaining one-third of the pre-sequence is sufficient to ensure transport of the partially processed pre-protein through the thylakoid membrane and into the lumen. A thylakoid-bound thiol protease is believed to be responsible for the removal of the final portion of the transit peptide.

Since the first demonstration of the presence of chloroplast DNA sequences outside the plastids, it has become clear that chloroplast genes may occur in mitochondria or nuclei either intact or in a rearranged state. Furthermore, there are genes of mitochondrial origin in chloroplast DNA.

REVIEW QUESTIONS

1. Describe the structure of chloroplast DNA with an example.

2. Describe the special features of chloroplast genome.

3. Explain in detail the transcription and translation of chloroplast genes.

4. Describe the chloroplast genes which have been mapped on chloroplast DNA.

5. Discuss about the coordination between chloroplast genome and nuclear genome in the expression of chloroplast proteins.

6. Name some of the plastid functions encoded by nuclear genes.

7. Explain with an example, the import of proteins into chloroplasts.

3

MITOCHONDRION AND ITS GENOME

In cell biology, a **mitochondrion** (plural **mitochondria**; in Greek *mitos* = thread + *khondrion* = granule) is a membrane-enclosed organelle, found in most eukaryotic cells (Figure 3.1). Mitochondria are sometimes described as "cellular power plants," because they convert food molecules into energy in the form of ATP via the process of oxidative phosphorylation. A typical eukaryotic cell contains about 2,000 mitochondria, which occupy roughly one-fifth of its total volume. Mitochondria also contain DNA that is independent of the nuclear DNA. According to the endosymbiotic theory, mitochondria are descended from free-living prokaryotes.

MITOCHONDRION STRUCTURE

Mitochondria are double-membrane-enclosed organelles. Mitochondria vary in shape from spherical to tubular. The size ranges from 1–10 micrometres (μm). The mitochondria have a smooth outer membrane and a highly convoluted inner membrane composed of phospholipid bilayers and proteins. The infoldings of the inner membrane are called **cristae**.

There are 5 distinct compartments (Figure 3.1) within mitochondria. They are (i) the outer membrane, (ii) the intermembrane space (the space between the outer and inner membranes), (iii) the inner membrane, (iv) the cristae, and (v) the mitochondrial matrix (space within the inner membrane).

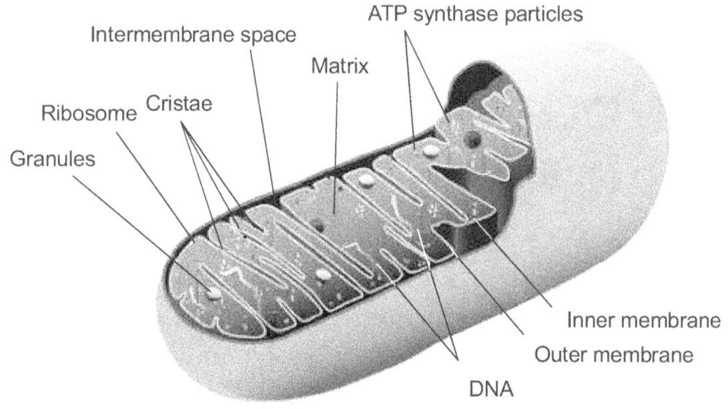

Figure 3.1 Simplified structure of a typical mitochondrion

Outer Membrane

The outer mitochondrial membrane, which encloses the entire organelle, has a protein-to-phospholipid ratio similar to the eukaryotic plasma membrane (about 1 : 1 by weight). It contains numerous integral proteins called porins, which contain a relatively large internal channel (about 2–3 nm) that is permeable to all molecules of 5000 daltons or less. Larger molecules can traverse the outer membrane only by active transport. It also contains enzymes involved in such diverse activities as the elongation of fatty acids, oxidation of epinephrine (adrenaline), and the degradation of tryptophan.

Inner Membrane

In contrast to the mitochondrial outer membrane and all other membranes in the cell, 70% of the inner membrane of a mitochondrion is composed of proteins. The inner membrane contains some unique phospholipids such as cardiolipins. The proteins in and on the inner membrane have special enzymatic and transport capabilities. The inner mitochondrial membrane contains proteins with four types of functions.

1. Proteins that carry out the oxidation reactions of the respiratory chain.

2. Proteins that make ATP in the matrix, e.g. ATP synthase.

3. Specific transport proteins that regulate the passage of metabolites into and out of the matrix.

4. Proteins that import nuclear-DNA-encoded mitochondrial proteins.

It contains more than 100 different polypeptides, and has a very high protein-to-phospholipid ratio (more than 3:1 by weight, which is about 1 protein for 15 phospholipids). Unlike the outer membrane, the inner membrane does not contain porins, and is highly impermeable to the passage of hydrogen ions (H^+); almost all ions and molecules require special membrane transporters to enter or exit the matrix. In addition, there is a membrane potential across the inner membrane. In other words, it serves as a barrier to the movement of protons. This important feature allows the formation of electrochemical gradients (gradient refers to the difference in the concentration of H^+ ions between the two sides of the inner mitochondrial membrane). The dissipation of the above-mentioned gradient is coupled to the formation of ATP by the phosphorylation of ADP that is mediated by the ATP synthase. ATP thus formed is released to other cellular sites where energy is needed to drive specific functions. The mitochondrial cristae expand the surface area of the inner mitochondrial membrane, enhancing its ability to generate ATP.

Mitochondrial Matrix

The matrix is the space enclosed by the inner membrane. It contains a highly concentrated mixture of hundreds of enzymes, in addition to the special mitochondrial ribosomes, tRNAs, and several copies of the mitochondrial DNA genome. Of these, the enzymes involved in the citric acid cycle and in the oxidation of pyruvate are the most important.

MITOCHONDRIAL ORIGIN

Mitochondria (as well as chloroplasts) are believed to have arisen from prokaryotic symbionts and many observations strengthen this theory. Mitochondrial DNA and a number of aspects of mitochondrial gene expression resemble those of bacteria. In particular, protein synthesis that occurs inside the mitochondria is not inhibited by cycloheximide, which inhibits the eukaryotic protein synthesis but not the bacterial protein synthesis. It is widely believed that prokaryotic cells were the first to achieve aerobic metabolism and that symbiotic incorporation of such bacteria conferred aerobic capacity on early eukaryotic organisms. Even today, mitochondria are the organelles solely responsible for aerobic metabolism.

MITOCHONDRIAL FUNCTION

They play a major role in the respiration process where the energy released after the metabolism of glucose is used for the synthesis of ATP. Hence, they are the cellular sites of respiration. Cellular respiration is the main route by which chemical energy is harvested from food and is converted to ATP. This process makes use of oxygen and glucose to produce carbon dioxide, water, and energy in the form of ATP and so is also known as **aerobic respiration**.

In photosynthesis the energy present in sunlight is changed to a chemical energy (carbohydrate) with the help of the ATP produced during the light reactions. In the cellular respiration process, this chemical energy (carbohydrate) is transformed to a form (ATP) that can be used by the organism. Adenosine triphosphate (ATP) is the energy-currency of the cell and it is used for driving the energy-consuming reactions of the organism. The differences between aerobic respiration and photosynthesis are given in Table below.

Aerobic Respiration	Photosynthesis
Energy-releasing pathway	Energy-storing pathway
Requires oxygen	Releases oxygen
Releases carbon dioxide	Requires carbon dioxide

The pathways involved in cellular respiration include:

✗ Glycolysis or splitting of sugar, which involves the oxidation of glucose to pyruvate with release of ATP

✗ Pyruvate oxidation, also known as the transition reaction

✗ Krebs cycle

✗ Electron transport chain /oxidative phosphorylation/ chemiosmosis

Glycolysis occurs in cytosol, just outside the mitochondria. The migration of compounds from cytosol to matrix should occur because the other three processes occur inside the mitochondria. Figure 3.2 shows the four stages involved in the cellular respiration.

Therefore, it is clear that aerobic respiration is the biological process by which reduced organic compounds are mobilized and subsequently oxidized. During respiration, free energy is released and transiently stored in a compound (ATP), which can be readily utilized for the maintenance and development of the organism.

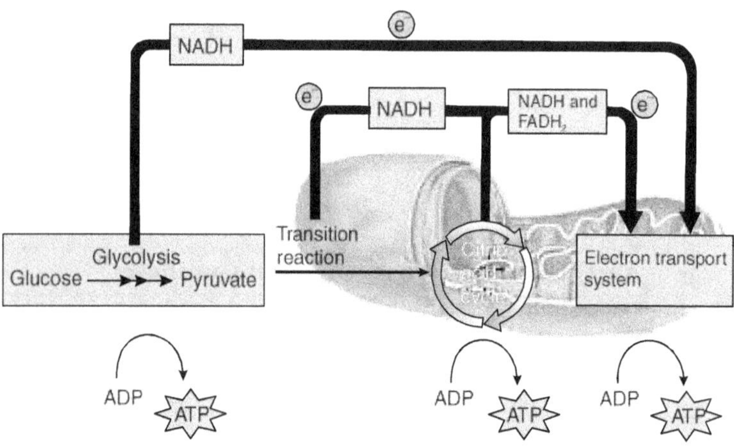

Figure 3.2 Cellular respiration showing the four stages

Glucose is the most commonly cited substrate for respiration. However, in a functioning plant cell, the reduced carbon is derived from sources such as the disaccharide sucrose, hexose phosphates and triose phosphates (Figure 3.3a) from starch degradation and photosynthesis (Figure 3.3b), fructose-containing polymers (fructans) and other sugars, as well as lipids (primarily triacylglycerols), organic acids and on occasion, proteins.

The carbohydrates synthesized by photosynthesis in the chloroplast are oxidized partially in the cytosol by glycolysis, and complete oxidation back to CO_2 and H_2O occurs in the mitochondria, releasing ATP in the process (Figure 3.3c). The pentose-phosphate pathway in the cytosol provides the NADPH necessary for the anabolic pathways and ribose sugars for nucleic acid synthesis.

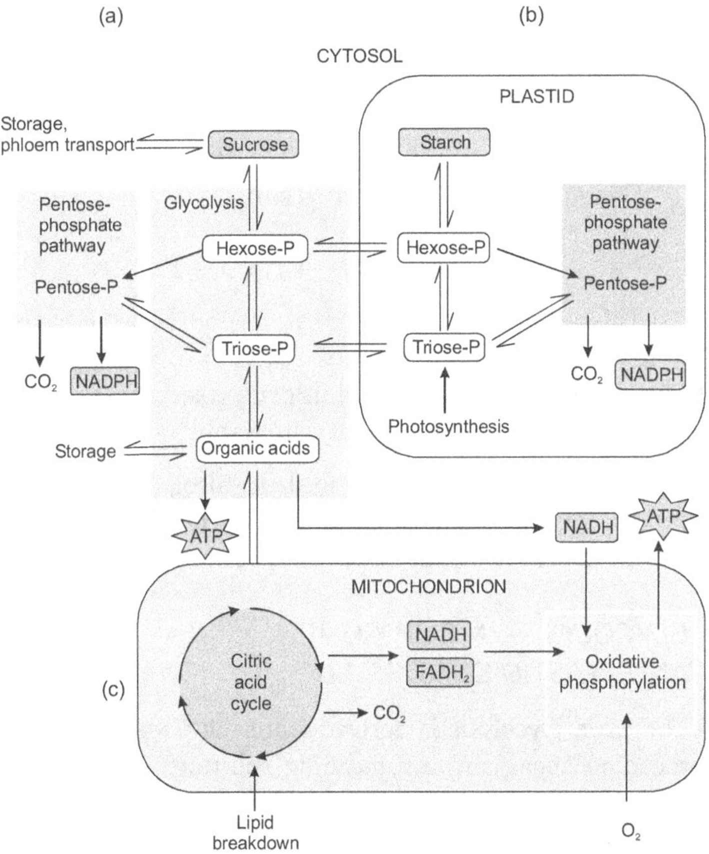

Figure 3.3   An overview of the various metabolic pathways that occur in different compartments of the plant cell. (a) glycolysis and pentose-phosphate pathway occurring in the cytosol (b) photosynthesis and production of starch occurring in the plastid or chloroplast (c) cellular respiration including the citric acid cycle and oxidative phosphorylation occurring in the mitochondria.

An overview of metabolism is given in Figure 3.3. From a chemical standpoint, plant respiration can be expressed as the oxidation of the 12-C molecule sucrose and the reduction of 12 molecules of O_2:

$$C_{12}H_{22}O_{11} + 13H_2O \rightarrow 12CO_2 + 48e^-$$

$$12O_2 + 48H^+ + 48e^- \rightarrow 24H_2O$$

Giving the following net reaction:

$$C_{12}H_{22}O_{11} + 12O_2 \rightarrow 12CO_2 + 11H_2O$$

This reaction is the reversal of the photosynthetic process. It represents a coupled redox reaction in which sucrose is completely oxidized to CO_2 while oxygen serves as the ultimate electron acceptor, and is reduced to water.

Though glycolysis occurs in the cytosol, an idea about the process is necessary and is discussed in the following section in order to obtain a complete picture of cellular respiration.

GLYCOLYSIS—A CYTOSOLIC AND PLASTIDIC PROCESS

The word glycolysis is derived from the Greek words, *glykos* meaning "sugar" and *lysis*, meaning "splitting". During glycolysis, carbohydrates such as glucose are converted to hexose phosphates that are then split into two triose (3-C) phosphates. As the foundation of both aerobic and anaerobic respiration, glycolysis is one of the most universal metabolic processes known and occurs (with variations) in many types of cells in nearly all organisms. Glucose is oxidized to either pyruvate or lactate depending on the aerobic or anaerobic mode respectively.

Aerobic glycolysis of glucose to pyruvate requires two equivalents of ATP to activate the process, with the subsequent production of four equivalents of ATP and two equivalents of NADH. Thus, conversion of one mole of glucose to two moles of pyruvate is accompanied by the net production of two moles each of ATP and NADH.

When oxygen is depleted, as for instance during prolonged vigorous exercise, the dominant glycolytic product in many tissues is lactate and the process is known as **anaerobic glycolysis**. Glycolysis, through anaerobic respiration, is the main energy source in many prokaryotes, eukaryotic cells devoid of mitochondria (e.g. mature erythrocytes) and eukaryotic cells under low oxygen conditions (e.g. heavily exercising muscle or fermenting yeast).

Glycolysis is the initial process of most carbohydrate catabolism, and it serves three principal functions.

1.   Generation of high-energy molecules (ATP and NADH).

2.   Production of pyruvate for the citric acid cycle.

3.   Production of various six- and three-carbon intermediate compounds, which may be used for other cellular processes.

The most common and well-known type of glycolysis is the Embden–Meyerhoff pathway, initially elucidated by Gustav Embden and Otto Meyerhoff. However, glycolysis will be used here as a synonym for the Embden–Meyerhoff pathway. The pathway has two stages, namely **preparatory phase and pay-off phase**, each having a set of five reactions. Thus, ten steps are involved in the production of glucose to pyruvate.

Priming stage (preparatory phase)   The preparatory phase includes the first five set of reactions, which prepares the sugar for energy production and here, no ATP production is involved. In this stage, two ATP molecules are consumed in order to break down a molecule of glucose (Figure 3.4). Glucose is converted to glucose 6-phosphate catalysed by the enzyme hexokinase. This is the first priming reaction where a molecule of ATP is consumed. The glucose 6-phosphate (aldose sugar) is converted to its isomer fructose 6-phosphate (ketose sugar).

The fructose 6-phosphate is converted to fructose 1,6-bisphosphate. The enzyme catalysing this reaction is phosphofructokinase. This is the second priming reaction where again a molecule of ATP is consumed. The six-carbon compound is cleaved to give two three-carbon intermediates: dihydroxy acetone phosphate and glyceraldehyde 3-phosphate, wherein, dihydroxy acetone phosphate is isomerized to glyceraldehyde 3-phosphate. Glyceraldehyde 3-phosphates serve as the substrate for the following reactions.

Pay-off stage (*energy-generation phase*)   It is the phase in which the high-energy compounds such as ATP and NADH are synthesized. It starts with a reaction that yields an NADH by the conversion of glyceraldehyde 3-phosphate to 1,3-bisphosphoglycerate catalysed by the enzyme glyceraldehyde 3-phosphate dehydrogenase. The reaction following this conversion is a substrate-level phosphorylation where an ATP molecule is synthesized. Here, 1,3 bisphosphoglycerate is converted to 3-phosphoglycerate by 1,3-bisphosphoglycerate kinase. This is the first ATP-forming reaction (Figure 3.4). 3-phosphoglycerate is converted to 2-phosphoglycerate by the enzyme phosphoglycerate mutase. The intermediate compound phosphoenol pyruvate (PEP) is synthesized from 2-phosphoglycerate by the enzyme 2-phosphoglycerate kinase. PEP is finally converted to pyruvate with the production of a molecule of ATP by the enzyme PEP kinase. This is the second ATP-forming reaction as shown in Figure 3.4.

Altogether in this phase, 4 equivalents of ATP (2 ATP from each of the trioses, namely glyceraldehyde 3-phosphate and dihydroxyacetone phosphate) and 2 equivalents of NADH are synthesized.

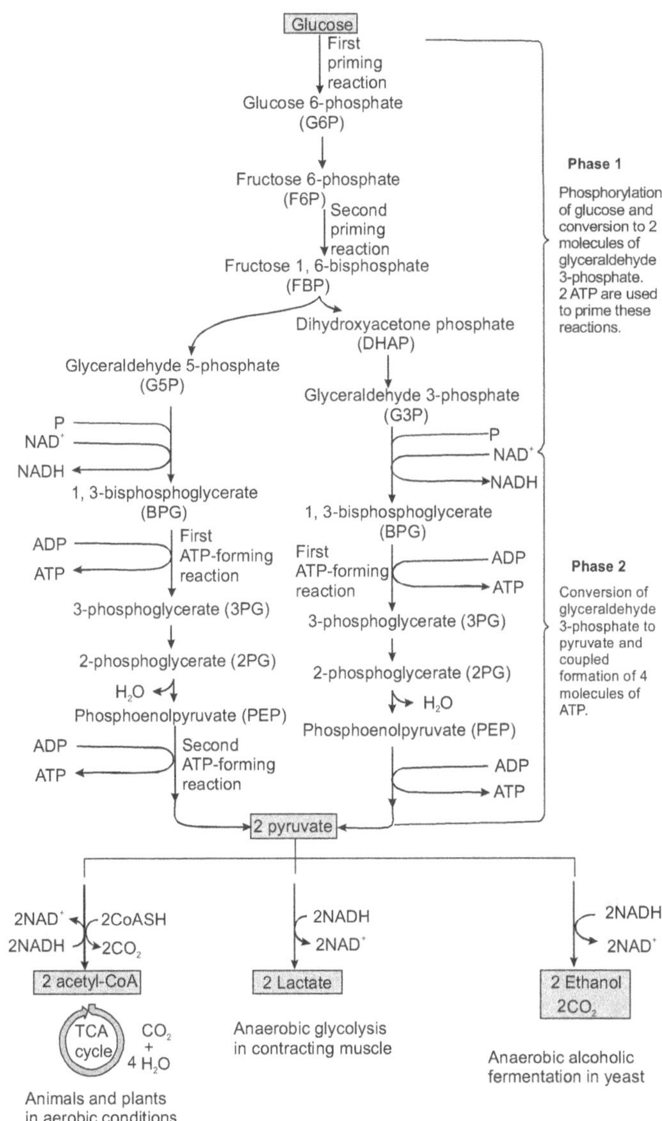

$$\text{Glucose} + 2\,\text{Pi} + 2\,\text{ADP} + 2\,\text{NAD} \longrightarrow 2\,\text{Pyruvate} + 2\,\text{H}_2\text{O} + 2\,\text{ATP} + 2\,\text{NADH} + 2\,\text{H}^+$$

Figure 3.4 Glycolysis pathway

Pentose-Phosphate Pathway (PPP)

In most microbes, the EMP is the major pathway for glucose conversion to pyruvate. But the EMP pathway neither provides the five-carbon sugars required for DNA and RNA formation nor erythrose 4-phosphate, a critical intermediate for aromatic amino acid synthesis. The glycolytic pathway is not the only route available for the oxidation of sugars in plant cells. Sharing common metabolites, the oxidative pentose phosphate pathway (also known as the **hexose monophosphate shunt**) can also accomplish this task. This pathway produces NADPH and biosynthetic intermediates. The reactions are carried out by soluble enzymes present in the cytosol and in plastids (cytosolic process). Generally, the pathway in plastids predominates over the cytosolic pathway (Dennis *et al.*, 1997). The oxidative pentose-phosphate pathway plays several roles in plant metabolism.

- ✖ The product of the two oxidative steps is NADPH, which is thought to drive reductive steps associated with various biosynthetic reactions that occur in the cytosol. In nongreen plastids such as amyloplasts, and in chloroplasts functioning in the dark, the pathway may also supply NADPH for biosynthetic reactions such as lipid biosynthesis and nitrogen assimilation.

- ✖ Because plant mitochondria are able to oxidize cytosolic NADPH via an NADPH dehydrogenase localized on the external surface of the inner membrane, some of the reducing power generated by this pathway may contribute to cellular energy metabolism, i.e., electrons from NADPH may end up reducing O_2 and generating ATP.

- ✖ The pathway produces ribose 5-phosphate, a precursor of the ribose and deoxyribose that is needed for the synthesis of RNA and DNA.

✗ Another intermediate in this pathway, the four-carbon erythrose 4-phosphate, combines with PEP in the initial reaction that produces plant phenolic compounds, including the aromatic amino acids and the precursors of lignin, flavonoids, and phytoalexins.

✗ During the early stages of greening, before leaf tissues become fully photoautotrophic, the oxidative pentose phosphate pathway is thought to be involved in generating Calvin cycle intermediates.

The two pathways glycolysis and PPP are extremely linked. They share common intermediates glyceraldehyde 3-P, fructose 6-P and glucose 6-P. The principal function of this pathway is to generate the reduced cofactor NADPH.

This pathway also has two phases, namely, oxidative phase and non-oxidative phase (Figure 3.5).

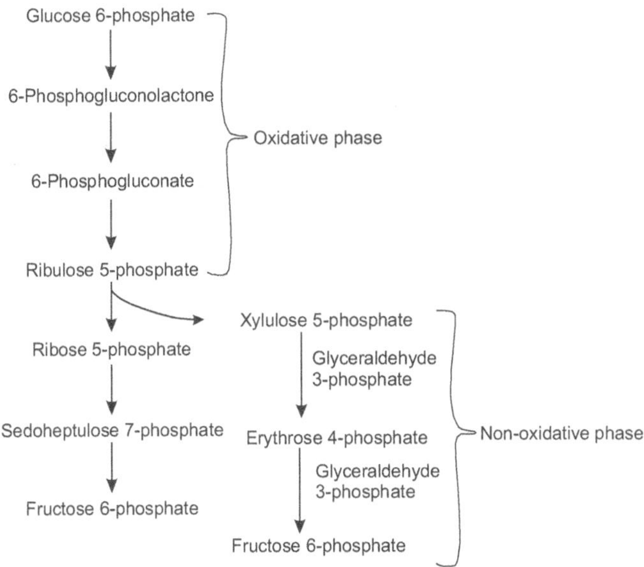

Figure 3.5 Pentose-phosphate pathway showing oxidative and non-oxidative phases

Oxidative Phase

The oxidative phase is the phase in which NADPH is generated. It involves oxidative events that convert the six-carbon glucose 6-phosphate to a five-carbon sugar, ribulose 5-phosphate. It is carried out in three steps.

Step I: Glucose 6-phosphate dehydrogenase catalyses the oxidation of the aldehyde (hemiacetal), at C1 of glucose 6-phosphate, to a carboxylic acid in ester linkage (lactone). **NADP⁺** serves as electron acceptor.

Step II: 6-Phosphogluconolactonase catalyses the hydrolysis of 6-phosphogluconolactone to 6-phosphogluconate.

Step III: Phosphogluconate dehydrogenase catalyses oxidative decarboxylation of 6-phosphogluconate, to yield the

5-C ketose ribulose 5-phosphate. This promotes loss of the carboxyl group at C1 as CO_2. $NADP^+$ again serves as an oxidant (electron acceptor).

Non-oxidative Phase

The non-oxidative phase generates ribose precursors. The reactions in this phase convert the five-carbon sugar (ribulose 5-phosphate) to the glycolytic intermediates such as glyceraldehyde 3-phosphate and fructose 6-phosphate. The steps involved in this phase are as follows:

The ribulose 5-phosphate formed in the oxidative phase acts as a substrate for two enzymes: epimerase and isomerase. The enzyme epimerase interconverts the stereoisomers ribulose 5-phosphate and xylulose 5-phosphate, and isomerase converts the ketose ribulose 5-phosphate to the aldose ribose 5-phosphate.

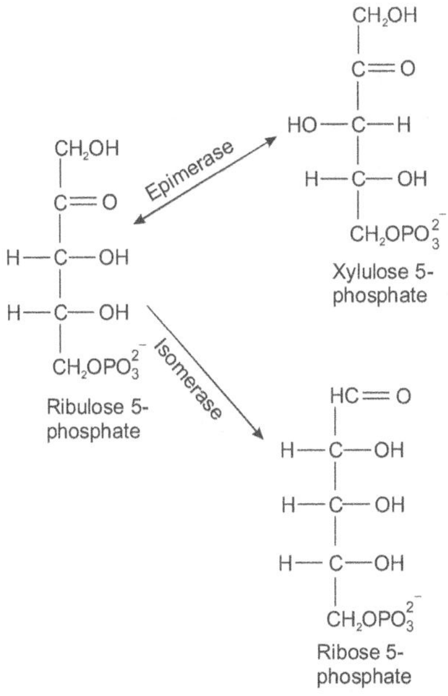

Transketolase transfers a 2-C fragment from xylulose 5-phosphate to ribose 5-phosphate. Therefore a seven-carbon compound namely sedoheptulose 7-phosphate is produced. Two five-carbon compounds react to produce a seven-carbon compound $(C_5 + C_5 \xrightarrow{\text{transketolase}} C_3 + C_7)$.

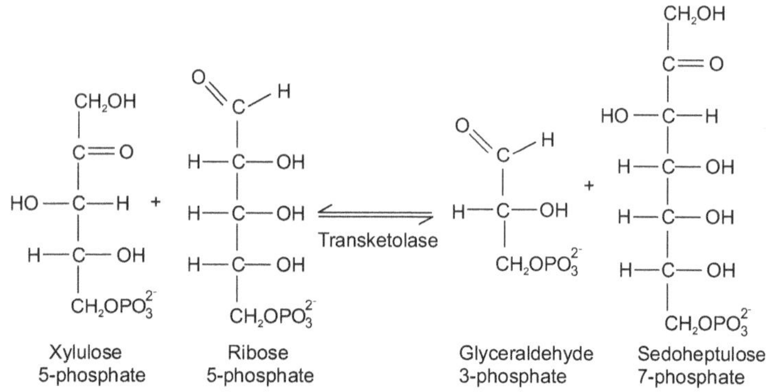

Xylulose 5-phosphate   Ribose 5-phosphate   Glyceraldehyde 3-phosphate   Sedoheptulose 7-phosphate

Also, transketolase catalyses the reaction between the five-carbon compound xylulose 5-phosphate and a four-carbon compound erythrose 4-phosphate to form a three-carbon compound and a six-carbon compound fructose 6-phosphate $(C_5 + C_4 \xrightarrow{\text{transketolase}} C_6 + C_3)$.

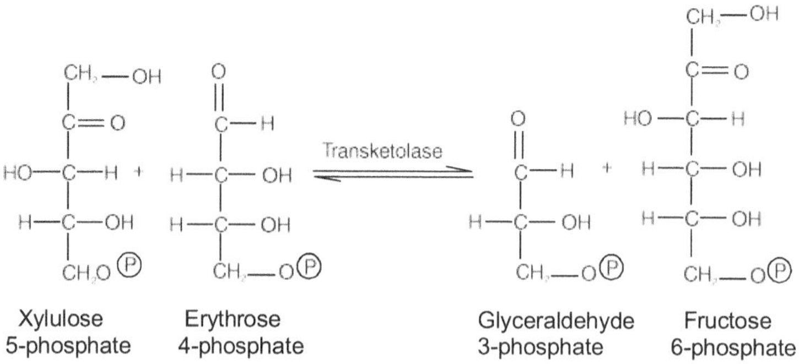

Xylulose 5-phosphate   Erythrose 4-phosphate   Glyceraldehyde 3-phosphate   Fructose 6-phosphate

Transaldolase catalyses the transfer of a 3-C dihydroxyacetone moiety, from sedoheptulose 7-phosphate to glyceraldehyde 3-phosphate. In other words, the seven-carbon compound reacts with the three-carbon compound in the presence of transaldolase to form a six-carbon compound and a four-carbon compound $(C_3 + C_7 \xrightarrow{\text{transaldolase}} C_6 + C_4)$.

Sedoheptulose 7-phosphate + Glyceraldehyde 3-phosphate $\underset{\text{Transaldolase}}{\rightleftharpoons}$ Erythrose 4-phosphate + Fructose 6-phosphate

Glucose 6-phosphate may be regenerated from either the 3-C product glyceraldehyde 3-phosphate or the 6-C product fructose 6-phosphate, via enzymes of gluconeogenesis. The net result is the complete oxidation of one glucose 6-phosphate molecule to CO_2 with the concomitant synthesis of 12 NADPH molecules. Studies of the release of $14CO_2$ from isotopically labelled glucose indicate that glycolysis is the more dominant breakdown pathway, accounting for 80 to 95% of the total carbon flux in most plant tissues. However, the pentose phosphate pathway does contribute to the flux, and developmental studies indicate that its contribution increases as plant cells develop from a meristematic to a more differentiated state.

CITRIC ACID CYCLE—A MITOCHONDRIAL MATRIX PROCESS

In the presence of air, cells consume O_2 and produce CO_2 and H_2O. In 1937 the German-born British biochemist Hans A. Krebs reported the discovery of the **citric acid cycle**, also called the **tricarboxylic acid cycle, the TCA cycle** or **Krebs cycle**. The elucidation of the citric acid cycle not only explained how pyruvate is broken down to CO_2 and H_2O, but also highlighted the key concepts of cycles in metabolic pathways. For his discovery, Hans Krebs was awarded the Nobel Prize in physiology and medicine in 1953.

The citric acid cycle is a series of chemical reactions of central importance in all living cells that utilize oxygen as part of cellular respiration. In aerobic organisms, the citric acid cycle is part of a metabolic pathway involved in the chemical conversion of carbohydrates, fats and proteins into carbon dioxide and water to generate a form of usable energy. It takes place in the mitochondria. It is the second of three metabolic pathways that are involved in the catabolism of fuel molecules and in the production of ATP—the other two being glycolysis and oxidative phosphorylation.

The citric acid cycle is also known as the tricarboxylic acid cycle, because of the formation of the tricarboxylic acids, citric acid (citrate) and isocitric acid (isocitrate) as early intermediates (Figure 3.6). The citric acid cycle that occurs in the mitochondrion begins with the oxidation of pyruvate produced during glycolysis.

Once inside the mitochondrial matrix, pyruvate is decarboxylated in an oxidation reaction by the enzyme pyruvate dehydrogenase. The products are NADH (from NAD^+), CO_2, and acetic acid in the form of acetyl-CoA, in which a thioester bond links the acetic acid to a sulphur-containing cofactor, coenzyme A (CoA). Pyruvate dehydrogenase exists as a large complex of several enzymes that catalyse the overall reaction in a three-step process: decarboxylation, oxidation, and conjugation to CoA.

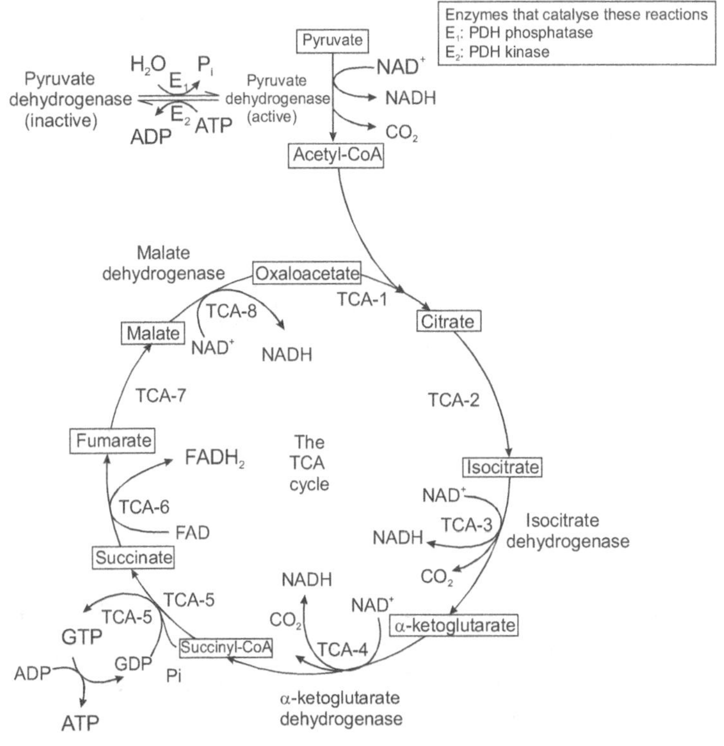

Figure 3.6   Citric acid cycle

In the next reaction, the enzyme citrate synthase combines the acetyl group of acetyl-CoA with a four-carbon dicarboxylic acid (oxaloacetate, OAA) to give a six-carbon tricarboxylic acid (citrate). Citrate is then isomerized to isocitrate by the enzyme aconitase.

The following two reactions are successive oxidative decarboxylation reactions the first catalysed by isocitrate dehydrogenase to produce α-ketoglutarate and the second catalysed by α-ketoglutarate dehydrogenase to produce succinyl-CoA. Each reaction produces one NADH and releases one molecule of CO_2, yielding a four-carbon molecule, succinyl-CoA. At this point, three molecules of CO_2 have been produced for each pyruvate that entered the mitochondrion, or twelve CO_2 for each molecule of sucrose oxidized. During the remainder of the citric acid cycle, succinyl-CoA is oxidized to oxaloacetic acid (OAA), allowing the continued operation of the cycle. Succinyl-CoA is converted to succinate by the enzyme succinyl-CoA synthetase, with simultaneous production of a molecule of GTP. The large amount of free energy available in the thioester bond of succinyl-CoA is conserved through the synthesis of ATP from ADP and Pi via a substrate-level phosphorylation catalysed by succinyl-CoA synthetase. (Recall that the free energy available in the thioester bond of acetyl-CoA was used to form a carbon–carbon bond in the step catalysed by citrate synthase.) The resulting succinate is oxidized to fumarate by succinate dehydrogenase, which is the only membrane-associated enzyme of the citric acid cycle and part of the electron transport chain (which is the next major topic to be discussed in this chapter) and reducing a molecule of FAD^+ to $FADH_2$. Fumarate is converted to malate by fumarase. Finally oxaloacetate (OAA) is produced by the action of malate dehydrogenase with reduction of NAD^+ to NADH. Most of the energy made available by the oxidative steps of the cycle is transferred as energy-rich electrons to NAD^+, forming NADH. For each acetyl group that enters the citric acid cycle, three molecules of NADH are produced.

The electrons and protons removed from succinate end up not on NAD^+ but on another cofactor involved in redox reactions, FAD (flavin adenine dinucleotide) that accepts electrons, forming $FADH_2$. FAD is covalently bound to the active site of succinate dehydrogenase and undergoes a reversible two-electron reduction to produce $FADH_2$ (Figure 3.7).

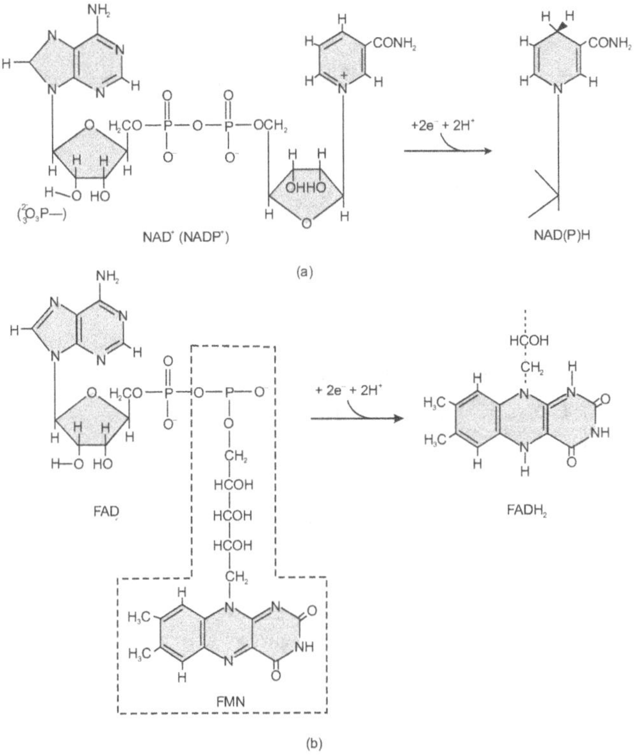

Figure 3.7 Structures and reactions of the major electron-carrying cofactors involved in respiratory bioenergetics. (a) Reduction of NAD(P)+ to NAD(P)H; (b) Reduction of FAD to FADH$_2$. FMN is identical to the flavin part of FAD and is shown in the dashed box.

At the end of each cycle, the four-carbon oxaloacetate is regenerated, and the cycle continues. Products of the first turn of the cycle are one ATP, three NADH, one FADH$_2$ and two CO$_2$.

The citric acid cycle also provides precursors for many compounds such as certain amino acids, and some of its reactions are therefore important. Because two acetyl-CoA molecules are produced from each glucose molecule, two cycles are required per glucose molecule.

OXIDATIVE PHOSPHORYLATION

In the citric acid cycle, we have seen that pyruvate from sugars and fatty acids are broken down to acetyl-CoA, and NAD^+ is reduced to NADH. In the process of oxidative phosphorylation, high-energy electrons from NADH that are produced during glycolysis, the pentose-phosphate pathway and the citric acid cycle are passed along the electron transport chain (also called respiratory chain) to oxygen in the inner mitochondrial membrane. This electron transfer releases a large amount of free energy, much of which is conserved through the synthesis of ATP from ADP and Pi (inorganic phosphate) catalysed by the enzyme ATP synthase. In other words, oxidative phosphorylation is the process by which the energy stored in NADH and $FADH_2$ is used to produce ATP. This final stage completes the oxidation of sucrose.

Two main steps are involved.

1. Oxidation—electron transport chain
2. Phosphorylation

Oxidation—Electron Transport Chain

Electrons are moved from molecules with low reduction potential (low affinity for electrons) to molecules with successively higher reduction potential (higher electron affinity). The hydrogen atoms covalently linked to NADH (or $FADH_2$) are first separated into protons and electrons. The electrons pass through a series of electron carriers in the inner membrane at several steps and the protons are not returned permanently until the electrons have reached the end of the electron transport chain. They are used to neutralize the negative charges created by the addition of the electrons to the oxygen molecule.

Complexes involved in the electron transport process The electron carriers of the respiratory chain are organized into membrane-embedded supramolecular complexes.

The respiratory chain consists of four membrane-bound complexes (named I to IV) in the inner mitochondrial membrane for electron transport from NADH to O_2. These four complexes are discussed in brief below. ATP synthase (Figure 3.8e), recently named complex V, is involved in the phosphorylation process and hence discussed under phosphorylation.

Complex I NADH-CoQ oxidoreductase It has over 20 subunits containing iron–sulphur clusters and a flavoprotein (coenzyme flavin mononucleotide, FMN). It is the main entry point of electrons from "internal" NADH, i.e., NADH produced in the matrix space during the oxidation of malate, pyruvate, oxoglutarate, isocitrate and glycine. It is involved in the transfer of electrons from NADH to ubiquinone (CoQ). This transfer is coupled to the translocation of H^+ from the matrix to the intermembrane space. This is therefore also called the first coupling site or site I of proton extrusion (Figure 3.8).

CoQ or ubiquinone is a small, hydrophobic (lipid-soluble) molecule and is freely diffusible in the inner mitochondrial membrane. Because of its ability to bind both electrons and protons, it plays a central role in coupling electron flow to proton movement.

Complex II succinate-CoQ oxidoreductase It contains a membrane-bound enzyme succinate dehydrogenase (a flavoprotein and an iron–sulphur protein), which is already encountered in the citric acid cycle, cytochrome *b* and two iron–sulphur proteins. Electrons from succinate enter the respiratory chain via complex II.

Complex III-cytochrome c reductase The components of this complex include cytochrome *b* (bH and bL), cytochrome *c*1 and several iron–sulphur proteins. It is responsible for the transfer of electrons from ubiquinol to cytochrome c. Coenzyme Q and cytochrome *c* are not part of the respiratory complexes but can move freely in the membrane.

The 'Q cycle' is thought to be responsible for the proton transfer. During this, two protons are transported for every electron passing through complex III. Thus complex III constitutes 'site 2' of proton extrusion. Cytochromes can carry electrons, but not hydrogens. This is another location where hydrogen ions leave the matrix.

Complex IV-cytochrome c oxidase　It has 10 subunits and contains cytochromes *a* and a_3, as well as two Cu^{2+} ions. It is the terminal oxidase of the pathway. This complex generates a proton-motive force by two entirely different processes.

1. It transfers electrons from cytochrome *c* to water. For each electron donated by cytochrome *c*, there is uptake of one H^+ from the matrix side to the opposite side to form water.

2. It acts as a proton pump extruding one H^+ per electron to the intermembrane space. This makes complex IV the third coupling site, site 3 of proton extrusion.

The steps in the electron transport process are the following.

1. *Transfer of electrons from NADH to CoQ*　The NADH formed during glycolysis and citric acid cycle as a result of the oxidation of glucose or other sugars are transported into the mitochondria and the electrons are routed into the electron transport chain via complex I to CoQ (Figure 3.8a). The reaction that takes place is given by

$$NADH + H^+ + CoQ \xrightarrow{\text{complex I}} NAD^+ + CoQH_2$$

2. *Transfer of electrons from $FADH_2$ to CoQ*　As an alternative to the first step, electrons are transferred from $FADH_2$ to CoQ

via complex II (succinate-CoQ reductase complex). (Figure 3.8b). This complex does not move H^+ ions. Oxidation of $FADH_2$ does not result in the movement of as many protons across the membrane, and hence produces less ATP.

$$Succinate + E - FAD \rightarrow Furmarate + E - FADH_2$$

$$E - FADH_2 + Fe - S_{ox} (complex\,II) \rightarrow E - FAD + Fe\text{-}S_{red}$$

$$(complex\,II) + 2H^+$$

$$Fe - S_{red} (complex\,II) + CoQ + 2H+ \rightarrow Fe - S_{ox}$$

$$(complex\,II) + CoQH_2$$

$$Succinate + CoQ \rightarrow Fumarate + CoQH_2$$

3. *Transfer of electrons from CoQ to cytochrome c*   The electrons from the reduced CoQ are transferred to the next electron carrier cytochrome *c* through complex III (CoQH$_2$–cyt *c* reductase complex). CoQ is converted back to its oxidized form and is again ready to accept electrons from complexes I or II (Figure 3.8c).

4. *Transfer of electrons from cytochrome c to oxygen*   Electrons from the reduced cytochrome *c* are transferred to oxygen through complex IV (cytochrome *c* oxidase complex) to form water (Figure 3.8d).

As seen above at complexes I, III and IV, the energy released during electron transfer is used to move H^+ ions from the matrix to the inner mitochondrial space, generating a gradient of protons across the membrane. In summary, the energy released from the electron transport chain is used to move protons across the inner mitochondrial membrane, thereby creating an electrochemical gradient that stores potential energy.

Phosphorylation   The dissipation of the electrochemical gradient formed during the electron transport process is coupled to the production of ATP. In this step, the protons move back across the membrane through ATP synthase (complex V)

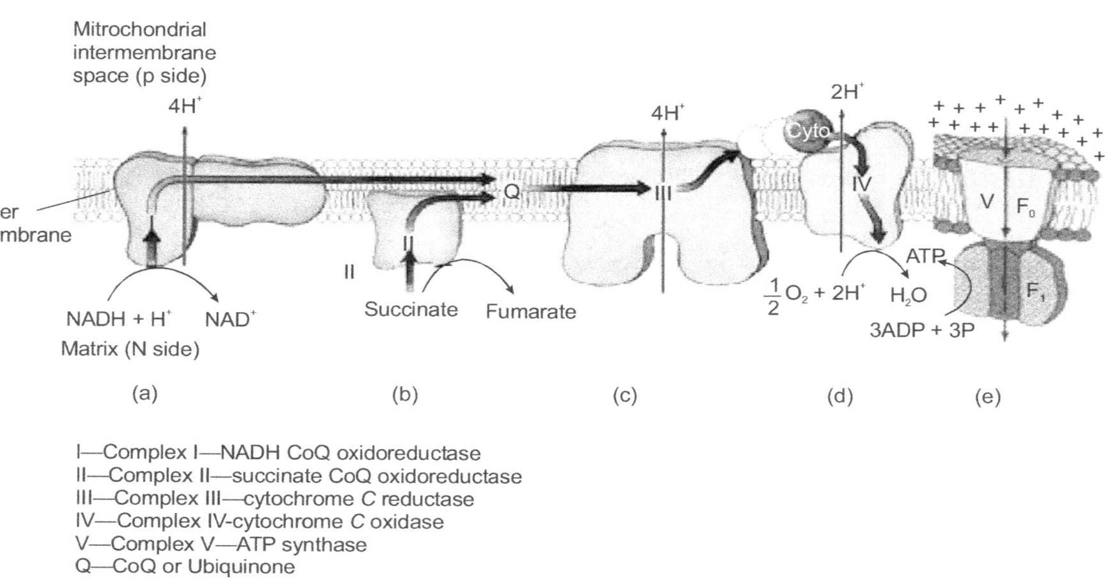

Figure 3.8　Steps involved in the electron transport process

providing the energy that is used to convert ADP to ATP (Figure 3.8e). The importance of the proton gradient in ATP production was first proposed by Peter Mitchell in 1961 and is called the chemiosmotic hypothesis: *the energy stored in an electrochemical gradient across the inner mitochondrial membrane could be coupled to ATP synthesis.* This hypothesis is now supported by substantial experimental data.

The enzyme that couples the proton motive force to ATP synthesis is the **F_0F_1 ATP synthase** (Figure 3.9). Movement of protons down the electrochemical gradient through a channel between the *a* and *c* subunits releases energy that is coupled to the rotation of the *c* and other subunits. This rotation causes a conformational change in the a and b subunits that promotes synthesis of ATP from ADP and Pi.

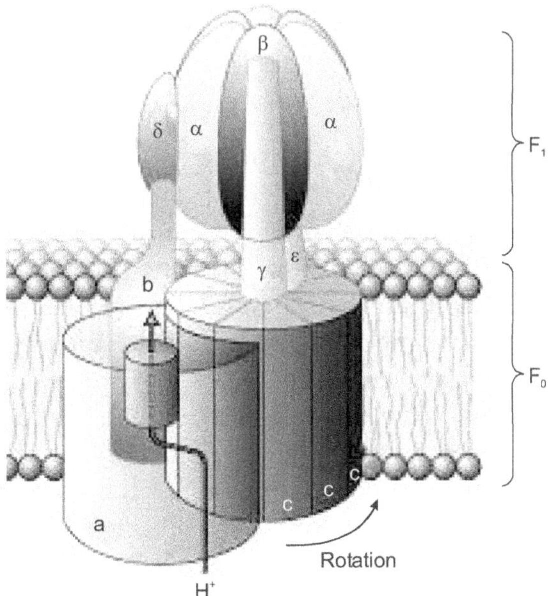

Figure 3.9   The ATP synthase complex

The other dehydrogenase is the rotenone-insensitive NADH dehydrogenase, which unlike complex I is not linked to proton extrusion.

External or cytosolic NAD(P)H feeds its electrons into the chain at the level of ubiquinone, similar to succinate. They are not connected with the translocation of H^+ across the inner mitochondrial membrane.

Other Factors Involved in Oxidative Phosphorylation

In addition to the proteins involved in the Krebs cycle, the electron transport chain, and ATP synthesis, a number of other factors are required for oxidative phosphorylation. Many of these are transporters and shuttles that move small molecules across the inner mitochondrial membrane. Transporters include the ATP/ADP antiporter, which transports ADP into the matrix and ATP out of the matrix. In addition there are separate transporters for Pi and pyruvate.

RESPIRATORY CONTROL

The rate of oxidative phosphorylation depends on the levels of ADP/ATP in mitochondria, i.e., oxidation of NADH and $FADH_2$ occurs only if there is ADP and Pi available to generate ATP. This phenomenon is termed **respiratory control**. The molecular mechanism of respiratory control involves the fact that the oxidation of NADH and $FADH_2$ is coupled to the transport of H^+ across the inner membrane. If the gradient of H^+ is not dissipated by the movement of H^+ through the F_0F_1 ATP synthase to make ATP from ADP and Pi, the gradient will become very steep. When this happens, $NADH/FADH_2$ oxidation (electron transport) will stop because the energy required for H^+ movement across the membrane will exceed the energy derived from electron transfer. There are also many other control points in respiration. For example, an increased level of ADP will stimulate glycolysis (described above).

Anaerobic Respiration

In the absence of oxygen, pyruvate is not metabolized by cellular respiration but undergoes fermentation. Pyruvate is not transported into the mitochondrion, but remains in the cytoplasm, where it is converted to waste products that may be removed from the cell. This waste product can vary depending on the organism. In muscles, the waste product is lactate, or lactic acid. In yeast, the waste product is ethanol and carbon dioxide.

2 ATP molecules per glucose are produced during anaerobic respiration, compared to 36 ATP per glucose produced by aerobic respiration (Table 3.1).

Table 3.1   An overall view of the mitochondrial metabolic activities

Step	Coenzyme yield	ATP yield	Source of ATP
Glycolysis (priming) preparatory phase		−2	Phosphorylation of glucose and fructose 6-phosphate uses two ATP from the cytoplasm.
Glycolysis pay-off phase		4	Substrate-level phosphorylation.
Glycolysis pay-off phase	2NADH	4	Oxidative phosphorylation. Only 2 ATP per NADH since the coenzyme must feed into the electron transport chain from the cytoplasm rather than the mitochondrial matrix.

(Contd.)

Table 3.1 (Continued)

Step	Coenzyme yield	ATP yield	Source of ATP
Transition reaction	2NADH	6	Oxidative phosphorylation.
Krebs cycle		2	Substrate-level phosphorylation.
Krebs cycle	6NADH	18	Oxidative phosphorylation.
Krebs cycle	2FADH$_2$	4	Oxidative phosphorylation.
Total yield		36 ATPs	From the complete oxidation of one glucose molecule to carbon dioxide and oxidation of all the reduced coenzymes.

PLANT MITOCHONDRIAL DNA (mtDNA)

Plant mitochondrial genome frequently occurs as a complex collection of multiple circular DNA molecules. Each mitochondrion contains multiple copies of the mitochondrial genome and a cell may contain many mitochondria. For example, a rat liver cell has 5 to 10 mtDNA molecules in each of about 1000 mitochondria. Thus each cell possesses from 5,000 to 10,000 copies of the mitochondrial genome. Like bacterial chromosomes, mtDNA lacks histone proteins.

SIZE AND COMPOSITION

It is generally circular and double-stranded. It has approximately 47% GC content and a buoyant density of about 1.706 g/mL in CsCl. The amount of mtDNA varies from 200–2,400 kb in different species of higher plants (Table 3.2).

Table 3.2 Mitochondrial DNA from different organisms

Organism	No. of kbp	No. of different molecules/ organelles
Higher plants		
Brassica sp.	218 kb	3 (circular) (218, 135, 83 kb)
Maize	570	7 (circular)
	+1.4 kb to 6 kb	4 (circular or linear)
Musk melon	2400	Not known
Fungi		
Podospora anserina	95	1 circular
Saccharomyces cerevisiae	80	1 circular
Others		
Cow and Man	16.6	1 circular

In maize, the main mitochondrial chromosome is a "master circle" of 570 kb (Figure 3.10).

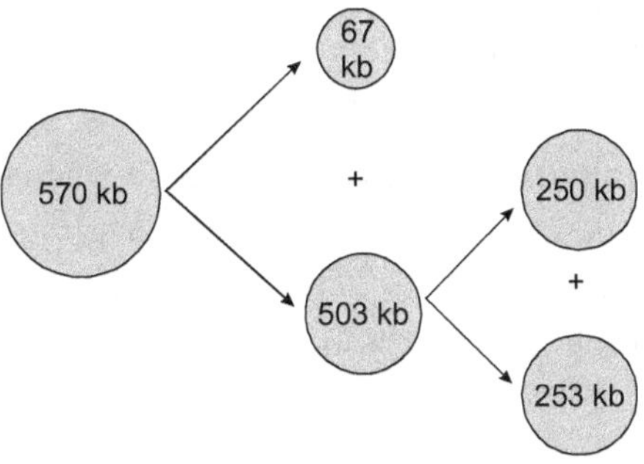

Figure 3.10 Recombination in the master circle of maize mitochondrial DNA

This contains six sets of repeated sequences, namely the 1, 2, 3, 10, 12 and 14 kb repeats and it has been proposed that recombination between them generates a range of smaller circular molecules. Recombination across the 12-kb repeats is believed to give rise to circles of 67 and 503 kb. Subsequent recombination across the 3-kb repeats generates circles of 250 and 253 kb.

MASTER CIRCLE AND SUBGENOMIC DNA MOLECULES OF FERTILE MAIZE

Mitochondria from several plants, including maize, sorghum and sugar beet contain small linear or circular DNA molecules in addition to the main chromosomal DNA. These can be detected by agarose gel electrophoresis of total mtDNA without digestion with restriction endonucleases.

Two linear DNAs called S1 (6.4 kb) and S2 (5.4 kb) occur in mitochondria of male sterile maize with S cytoplasm. Based on male sterility, cytoplasm of maize is divided into 4 general types: T (Texas), S (USDA), C (Charrua) and N (Normal) types. These two linear DNAs have identical 208-bp terminal inverted repeat sequences called S-TIRs. The presence of S1 and S2 is correlated with the synthesis of certain polypeptides of abnormal size in the mitochondria. DNA sequences homologous to the S-TIRs are found at various sites in the main mtDNA of maize lines with N and S cytoplasm.

Recombination across these sites between the linear S1 and S2 molecules and circular mtDNA gives rise to linear chromosomes with S-TIR sequences at their termini. Interestingly, these recombination events appear to be influenced by nuclear genes.

GENE STRUCTURE AND EXPRESSION

A number of functions have been assigned to plant mtDNA. The first sequences to be located on mtDNA were those encoding the RNA components of the 78S ribosomes, which are different from the corresponding rRNAs in the cytosol and chloroplasts. The plant mitochondrial 26S and 18S rRNAs are larger than their counterparts in animals and there is a unique 5S rRNA in higher plants (Table 3.3).

Table 3.3   Genes located on plant mitochondrial DNA

Genes for stable RNAs	26S rRNA
	18S rRNA
	5S rRNA
	tRNAs possibly 30
Protein-coding genes	NADH dehydrogenase(subunit 1)
	ATP synthase
	subunit A (*atp*A)
	subunit 6 (*atp*6)
	subunit 9 (*atp*9)
	Ubiquinol cytochrome *C* reductase
	Cytochrome *B* (CoB)
	Cytochrome *C* oxidase
	subunit I (*cox*I)
	subunit II (*cox*II)
	subunit III (*cox*III)
	Ribosomal proteins
	Small subunit
Chloroplast genes	3 genes for 16S rRNA
	For tRNA
	For the large subunit of RuBisCo

In maize, the 18S and 5S genes are closely linked, but are separated from the gene encoding 26S rRNA. Sequence analysis of the 3'-end of the 18S rRNA gene and the 5S rRNA gene show that they are transcribed from the same DNA strand and are separated by 108 bp (Chao *et al.*, 1983). The 26S coding sequence is approximately 16 kb away.

There is a sequence (5´-UGAAU-3´) in the 3´-end of the 18S rRNA which is believed to bind the 3´-ACUUA-5´ sequence found near the start of some plant mitochondrial mRNA sequences.

Comparison of the amino acid sequences with the DNA sequences of the corresponding genes shows the variations in the "universal" genetic code that occur. For plant mitochondria, the only variation appears to be the use of CGG to code for tryptophan in the place of arginine.

The presence of the gene for one of the subunits of ATP synthase (*atp*1) is very interesting, since in other organisms, including yeast and humans, it is encoded by the nucleus.

The major transcripts of the maize gene, revealed by northern blotting are only 4.2 and 2.2 kb long, suggesting that there may be some processing of RNA transcripts. In maize, the minimum length of a full transcript would be 1619 bases including the intron or 825 bases without the intron. Northern blotting has revealed a number of transcripts, the largest being about 3.5 kb. The results suggest that the transcripts may be spliced.

The sequence of the maize mitochondrial gene for subunit I of cytochrome oxidase encodes a polypeptide of 528 amino acids and contains no introns. The same gene in yeast mitochondria has 7–9 introns (depending on the strain) and the first 4 contain long open reading frames. Ten mutations have been introduced in the intron I of this gene and each of them has resulted in the failure to produce the cytochrome oxidase subunit I. This intron codes for the maturase or splicing enzyme involved in the processing of the mRNA precursors. Similarly the fourth intron in yeast mitochondrial gene for cytochrome *b* encodes a maturase involved in splicing of the same intron (class I intron).

Interestingly, this maturase also controls splicing of the fourth intron of the pre-mRNA for yeast **cytochrome oxidase subunit I**.

Promiscuous DNA

It is now apparent that DNA can move between organelles. This is called **promiscuous DNA**. For example, in maize, a 12-kb region homologous to chloroplast DNA has been shown to be inserted into the mtDNA (Stern and Lonsdale, 1982). The chloroplast genes in mtDNA include those for 16S rRNA, tRNA and the large subunit of RuBisCO. There is also evidence for the presence of a nuclear gene in the mitochondria of *Neurospora*, and seven genes similar to those of mitochondrial NADH dehydrogenase are present in tobacco chloroplast DNA. It is not yet clear how this transfer of DNA has occurred. It might have involved transposable elements or plasmid-like DNAs. The plasmid-like DNAs are known to occur in plant mitochondria and there is evidence that they can associate or dissociate from the mitochondrial chromosome.

MITOCHONDRIAL GENE EXPRESSION

Transcript Processing

The protein products of plant mitochondrial genes cannot be predicted accurately from genomic sequences since RNA editing modifies almost all mRNA sequences post-translationally. Typically, editing involves changes from C bases in the genome to U in the mRNA. RNA editing alters 5′ untranslated regions (UTRs), 3′ UTRs as well as introns and exons.

Several mitochondrial genes contain introns. The mitochondrial genome of the liverwort, *Marchantia polymorpha*, contains 32 self-splicing introns, 25 are members of group II introns and seven are classified as group I introns. However, no group I introns have been found in plant mtDNA.

In plant mitochondria, transcripts of the NADH dehydrogenase complex (*nad*1, 2, 5) are processed by *trans*-splicing that connects exons scattered throughout the genome (Figure 3.11).

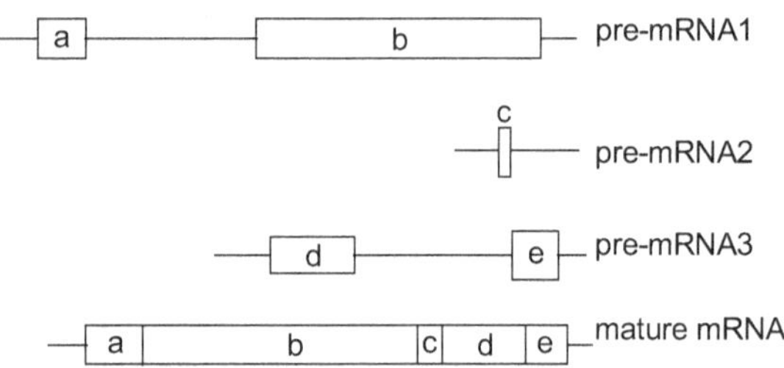

Figure 3.11   *Trans*-splicing assembly of the mature mRNA for the *nad*5 mitochondrial gene from three different pre-mRNA molecules involves both *trans*-and *cis*-splicing (Wissinger *et al.*, 1992).

TRANSLATION

The mitochondrial genome encodes part of the mitochondrial protein translation machinery. However, not all the required tRNA genes are present in mtDNA and additional tRNAs have to be imported into mitochondria from the cytoplasm.

The ribosomes of mitochondria are structurally more like prokaryotic ribosomes. They have a sedimentation coefficient of 78S and are inhibited by the antibiotics chloramphenicol and tetracycline. Cytoplasmic ribosomes (80S) are not inhibited by these antibiotics. Plant mitochondrial ribosomes contain a distinct 5S rRNA not found in other eukaryotic mitochondrial ribosomes. Like chloroplast mRNAs, those from mitochondria lack the 5′ cap and 3′ poly(A) tail.

IMPORT OF PROTEINS INTO MITOCHONDRIA

Most of the many hundreds of proteins present in mitochondria are encoded by nuclear genes, synthesized in the cytosol and imported subsequently. Protein transport into mitochondria has been studied well in yeast. It requires ATP and an energized inner membrane. The N-terminal pre-sequence directs the transport of protein into mitochondria. It determines the location of the protein in the outer or inner membranes, the intermembrane space or the matrix. Deletion of the pre-sequence impairs the transport of proteins into mitochondria.

Over 20 nuclear genes for mitochondrial proteins have been cloned and sequenced. There is no extensive homology between the pre-sequences of different proteins, although they share common characteristics.

Proteins located in the outer membrane lack cleavable pre-sequences, although they contain targeting information at their N-termini. For example, when the first 41 amino acids of the mature 70-kDa outer membrane protein was fused to β-glucosidase from *E. coli*, they directed this *E. coli* protein to the outer mitochondrial membrane, where it was inserted.

Another outer membrane protein called porin, forms diffusion pores for molecules with molecular weights below 6 kDa. It is encoded by nuclear genes, translated in the cytosol and inserted into the outer membrane of mitochondria post-translationally.

Proteins that are transported to the inner membrane and the matrix have a cleavable N-terminal pre-sequence. This region is generally rich in hydroxylated amino acids and is devoid of acidic residues. It contains one or more cleavage sites for proteases.

The function of the pre-sequence of the yeast alcohol dehydrogenase III, which is located in the mitochondrial matrix, has been confirmed. This pre-sequence was attached to the N-terminus of mouse dihydrofolate reductase, a cytosolic enzyme. The fused pre-sequence transported the protein to the matrix of the mitochondrion. The pre-sequence was unable to cross both the membranes and was cleaved on the membrane.

Location of some mitochondrial proteins	
Name of the protein	**Location in mitochondria**
Alcohol dehydrogenase III	Matrix
Subunit IV of cytochrome *c* oxidase	Inner membrane
Cytochrome *c*1	Intermembrane space
70-kDa membrane protein	Outer membrane

CYTOPLASMIC MALE STERILITY (CMS)

The inability to produce fertile pollens in many plants is controlled genetically. It results from pollen abortion at one of the several stages between meiosis and microspore mitosis. In all the plants examined so far, cytoplasmic male sterility is determined by genes present in the mitochondria. This sterility can be reverted to fertility by nuclear restorer (*Rf*) genes. CMS is a maternally inherited phenotype characterized by the inability of a plant to produce functional pollens. CMS can result from a wide variety of reproductive abnormalities. Many CMS loci result in failure of pollen development due to abnormalities in either the anther or the developing pollen (gametophytic) tissues. CMS phenotypes can be associated with abnormalities in anther vascular tissue, the tapetal cell layer that lines the anther locule, or the callose layer that surrounds the developing microspores.

Cytoplasmic male sterility has been extensively studied in maize where four general types of cytoplasm have been distinguished. They are **N** (normal fertile), **T** (Texas), **S** (US Department of Agriculture [USDA]) and **C** (Charrua). These types are distinguished from each other on the basis of the mode of restoration of fertility by nuclear *Rf* genes. Thus, genes *Rf*1 and *Rf*2 restore fertility to cms-T but not to cms-C or cms-S.

There is variation in the sequence organization, content and expression of mtDNA from fertile and male-sterile lines of maize. Mitochondria from male-sterile maize, carrying S cytoplasm, contain small linear DNA molecules, S1 and S2. These linear molecules encode for 9 additional polypeptides (Table 3.4).

In cms-T mtDNA, an open reading frame (ORF) encoding a 13-kDa protein (URF13) produces a transcript found only in male sterile plants (Figure 3.12). The gene producing URF13 (unidentified reading frame 13) is called T-*urf*13, originating from multiple rearrangements that have assembled the flanking and/or coding regions of the mitochondrial 26S rRNA gene (rrn26), the *atp*6 gene and a promiscuous chloroplast tRNA$_{arg}$ gene. This URF13 polypeptide is a component of the inner mitochondrial membrane.

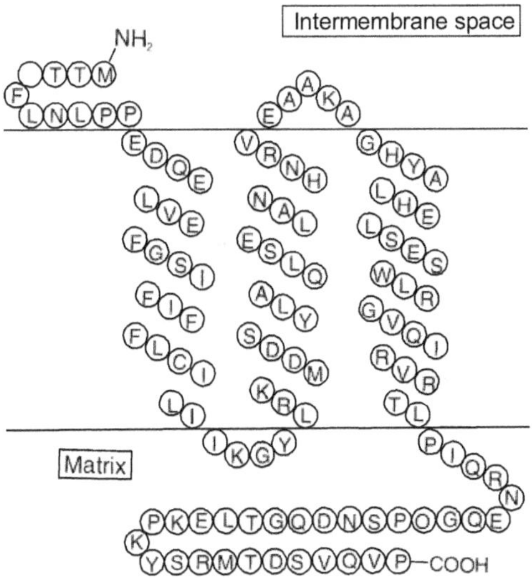

Figure 3.12 Model of the membrane location of the polypeptide URF13, associated with cytoplasmic male sterility (cms-T) in maize, showing the membrane-spanning helical domains (Levings and Siedow, 1992).

Table 3.4   Occurrence of low-molecular-weight DNA molecules in maize mitochondria

DNA length (kb)	Configuration	Cytoplasm			
		N	T	C	S
6.4	Linear	–	–	–	+ (S1)
5.4	Linear	–	☼	–	+ (S2)
2.35	–	+	–	+	+
1.94	Supercoiled circle	+	+	+	+
1.57	Circular	–	–	+	–
1.42	Circular	–	–	+	–

Although C and T male-sterile plants lack S1 and S2 molecules, they contain other low-molecular weight DNAs.

Occurrence of low-molecular-weight polypeptides in maize mitochondria is shown in the following table.

N	T	C
21 kDa	13 kDa	–
15.5 kDa	–	17.5 kDa

In the case of T mitochondria, a protein with a molecular weight of 13 kDa is synthesized. The corresponding protein in N (Normal) mitochondria is 21 kDa. A 15.5 kDa synthesized in N mitochondria is replaced by a larger protein of 17.5 kDa in C mitochondria.

Another example of an altered mitochondrial polypeptide associated with cytoplasmic male sterility has been demonstrated in *Sorghum* (Baily-Serres *et al.*, 1987). DNA sequence analysis shows that the normal *cox*I (*cox*I is the gene encoding cytochrome *C* oxidase) polypeptide is 38 kDa, whereas in the male-sterile lines, the protein is 42 kDa. Both *cox*I genes are identical at the 5′-end. The difference in size of the polypeptides is due to a recombination event that leads to an extension of the *cox*I gene at the 3′-end such that it encodes an aberrant polypeptide with an extra 101 amino acids at the C-terminus. Nuclear *Rf* genes appear to compensate for resorting fertility, although the aberrant protein is still made.

Chimeric Structure of Petunia CMS-associated DNA

A CMS-associated DNA (*S-pcf*) has been sequenced and found to be a mosaic or chimeric structure containing sequences from five different genes (Figure 3.13).

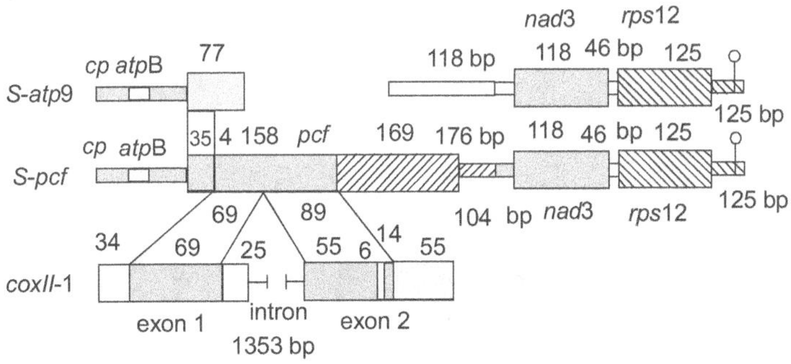

Figure 3.13 Mitochondrial DNA associated with cytoplasmic male sterility (CMS) in *Petunia* (the *S-pcf* locus) and homologous mitochondrial genes, showing the chimeric nature of this locus. Large boxes—Exons; Narrow boxes—Flanking regions; Size is indicated in bp

Transcripts and a protein product of the *S-pcf* gene have been identified in male-sterile plants. There is evidence to suggest that the nascent *S-pcf*-encoded polypeptide is post-translationally processed but the function of the protein is not known.

REVIEW QUESTIONS

1. Comment on the genome organization in mitochondria.

2. How are the proteins imported into mitochondria?

3. Comment on the chloroplast sequence in mitochondrial DNA.

4. Explain cytoplasmic male sterility and its significance.

5. Discuss the molecular basis of cytoplasmic male sterility. Add a note on the agricultural implications.

TRANSPOSABLE ELEMENTS

Transposable elements or transposons are specific DNA sequences that are transferred as a unit from one replicon to another. Based on the transposition mechanism, two distinct classes of transposable elements are identified. One class is structurally similar to transposable elements found in bacteria, typically ending in short inverted repeats (IRS) and transposing through DNA intermediates. The other class comprises retrotransposons in which, the elements move via an RNA intermediate; very few active retrotransposons have been identified in plants. The other class of transposons, move via direct excision and integration of DNA sequences, e.g. *Ac* and *Ds* elements of maize, and *Tam* 1, 2 and 3.

Ac AND *Ds* ELEMENTS IN MAIZE

About 50 years ago, Barbara McClintock reported transposable elements in maize (corn). She started her research on the maize genes that produced variegated (multicoloured) kernels. Most corn kernels are either wholly pigmented or colourless (yellow), but some yellow kernels had spots or streaks of red colour. The study of variegated corns led Barbara McClintock to discover transposable elements, for which she was awarded the Nobel Prize in physiology and medicine in 1983.

McClintock discovered that the reason for the unstable mutation was a jumping gene or a gene on the chromosomal DNA that moved from one region to the other. She noticed that breakage of chromosome in maize often occurred at a gene that she called **Dissociation** (*Ds*), which took place in the presence of another gene, called the **Activator** (*Ac*). *Ds* and *Ac* exhibited unusual patterns of inheritance. Occasionally, these genes moved together. McClintock called these moving genes as controlling elements because they controlled the expression of other genes.

The structure and function of *Ac* and *Ds* elements are similar to those of transposable elements found in bacteria. These are DNA transposons that possess terminal inverted repeats and generate flanking direct repeats at the points of insertion. Ac elements are about 4500 bp long, including terminal inverted repeats of 11 bp. The flanking direct repeats that they generate are 8 bp in length (Figure 4.1a).

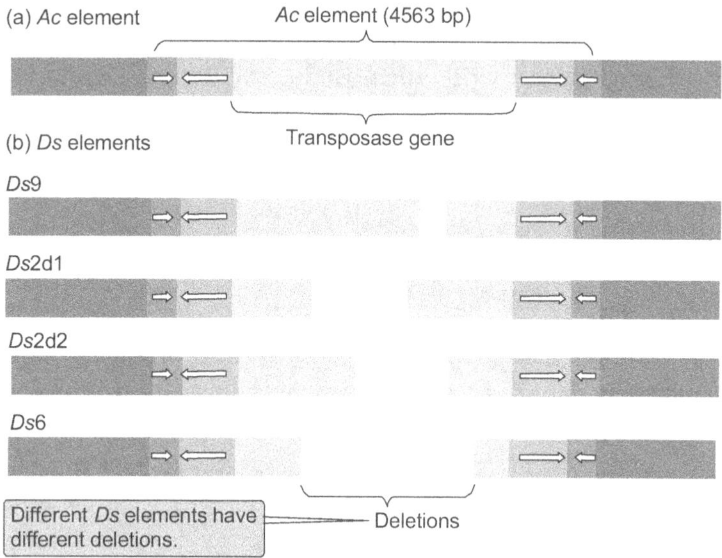

Figure 4.1   *Ac* and *Ds* transposable elements in maize

Each *Ac* element contains a single gene that codes for the enzyme transposase. Thus *Ac* elements are autonomous, i.e., they can transpose on their own. *Ds* elements are *Ac* elements with one or more deletions that have inactivated the transposase gene (Figure 4.1b). Unable to transpose on their own (non-autonomous), *Ds* elements can transpose in the presence of *Ac* elements because they still possess terminal inverted repeats that can be recognized by *Ac* transposase.

Each kernel in an ear of corn is a separate individual, originating as an ovule fertilized by a pollen grain. A kernel's pigment pattern is determined by several loci. A pigment-encoding allele at one of these loci can be designated *C*, and an allele at the same locus that does not confer pigment can be designated *c*. A kernel with genotype *cc* will be colourless, that is, yellow or white (Figure 4.2a); a kernel with genotype *CC* or *Cc* will produce pigment and be purple (Figure 4.2b).

A *Ds* element, transposing under the influence of a nearby *Ac* element, may get inserted into the *C* allele, thereby destroying its ability to produce pigment (Figure 4.2c). An allele that is inactivated by a transposable element is designated with a subscript "*t*"; so, in this case, it would be designated as C_t. After the transposition of *Ds* into the *C* allele, the kernel cell has genotype $C_t c$. This kernel will be colourless (white or yellow), because neither the C_t nor the *c* allele confers pigment.

As development takes place and the original one-celled maize embryo divides by mitosis, additional transpositions may take place in some cells. In any cell, in which the transposable element is excised from the C_t allele and is moved to a new location, the *C* allele is rendered functional again. All cells derived from those in which this event has taken place will have the genotype *Cc* and will be purple. The presence of these pigmented cells, surrounded by the colourless ($C_t c$) cells, produces a purple spot or streak (called a sector) in the otherwise yellow kernel (Figure 4.2d).

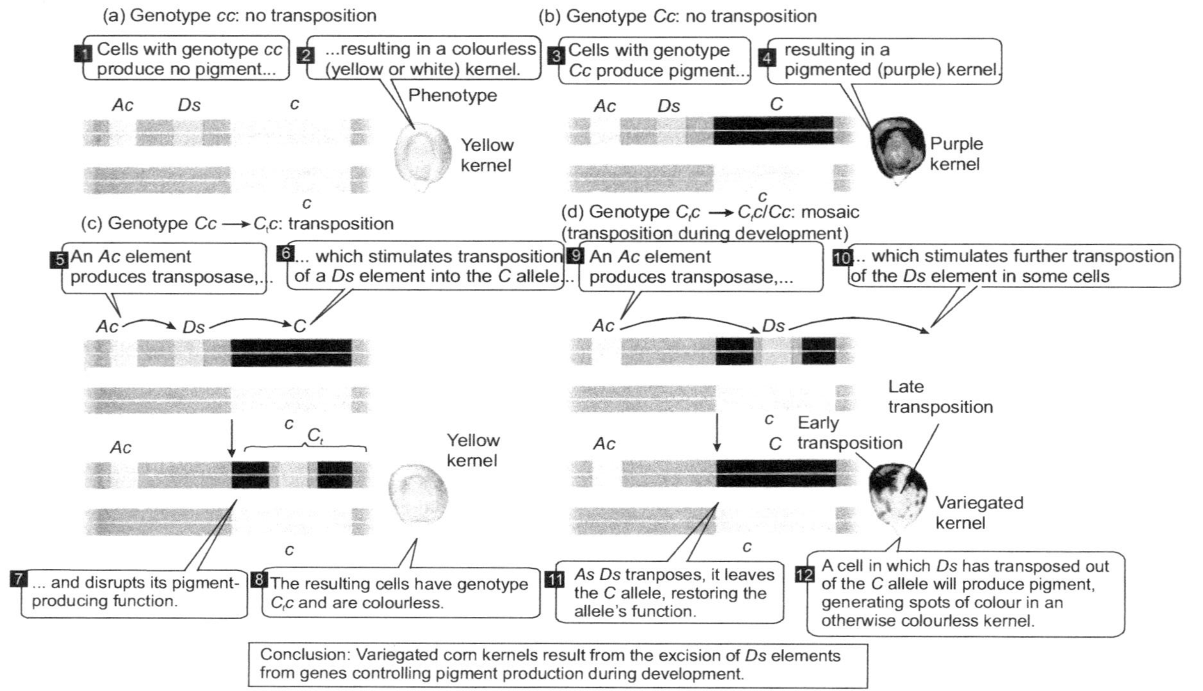

Figure 4.2 Transposition ending in variegated maize kernels

The size of the sector varies, depending on when the excision of the transposable element from the C_t allele occurred. If the excision occurred early in development, then many cells will contain the functional C allele and the pigmented sector will be large; if excision occurred late in development, few cells will have the functional C allele and the pigmented sector will be small.

Genotype	Phenotype
CC	Purple
cc	Colourless
Cc	Purple
C_tc	Yellow
$C_t(d)c$	Variegated kernel

TAM TRANSPOSONS IN *ANTIRRHINUM* FLOWER PIGMENTATION

The other classical example of transposition is the *Tam* transposon observed in *Antirrhinum*.

Tam transposons play a major role in flower pigmentation by the transposition of *Tam* elements in the biosynthetic pathway of anthocyanin, a pigment. *Tam1* and *Tam3* are autonomous transposable elements that can move in the genome and give rise to variegated (patterned) flowers whereas *Tam2* is a non-autonomous transposable element that cannot move in the genome by itself, and gives rise to a stable mutant phenotype. Figure 4.3 shows in outline the biosynthetic pathway for anthocyanin pigment synthesis and the positions of blocks caused by the *nivea* and *pallida* mutants.

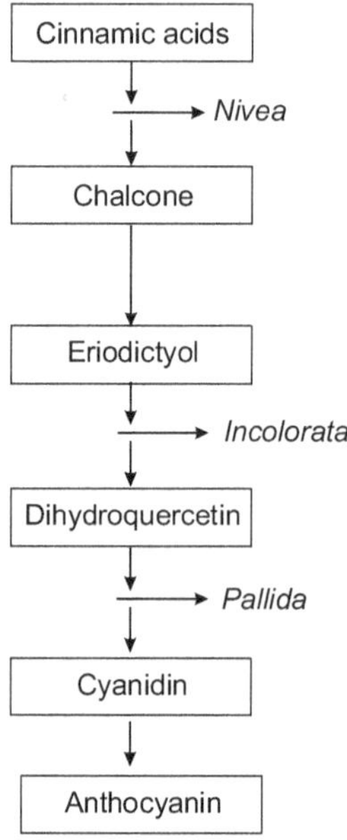

Figure 4.3   Biochemical pathway for anthocyanin biosynthesis in *Antirrhinum majus* (snapdragon) showing the positions of blocks caused by the *nivea* and *pallida* mutations. Arrows represent one or more enzyme steps.

Table 4.1 lists the transposable elements in *Antirrhinum majus* and its resulting phenotype.

Table 4.1   Transposable elements in *Antirrhinum majus*

Allele	Flower phenotype	Element
niv rec-53	Red sites on white background	*Tam*1
niv-44	Stable white	*Tam*2
niv rec-98	Red sites on pale background	*Tam*3
pal rec-22	Red sites on ivory background	*Tam*3

Table 4.2 shows a comparison of the *Ac* transposable element from the maize (*Zea mays*) with two types of transposable elements one from snapdragon (*Antirrhinum majus*) and another, a different, maize transposable element system (*En/Spm*-enhancer/suppressor) isolated by Barbara McClintock.

Ac and *Tam3* occur in different plants but they share certain features like IRs and target site duplication. Similarly, the *En/Spm* and *Tam1* of *Zea mays* and *A. majus* share certain features. *Tam3* from *A. majus* which is about the same size as *Ac*, has a 12-bp IR at the ends and contains a single ORF (with no introns). A region of 520 amino acids of the putative protein encoded by this ORF has 30% identity to the Ac transposase. A further similarity between *Ac* and *Tam3* is that both generate an 8-bp target site duplication when they are inserted into DNA.

Table 4.2   Comparison of plant transposons

Element	Species	Size (bp)	Terminal inverted repeat	Target site duplication
Ac	*Zea mays*	4563	CTAGGGATGAAA	8
Tam3	*A. majus*	4869	TAAAGATGTGAA	8
En/Spm	*Zea mays*	8287	CACTACAAGAAAA	3
Tam1	*A. majus*	15164	CACTACAAGAAAA	3

Table 4.2 also shows the structural similarities of the other two transposons with each other. Both *En/Spm* from maize and *Tam1* from *A. majus* produce a 3-bp target site duplication upon insertion. Although *En/Spm* and *Tam1* differ in size, they both are much larger than *Ac* and *Tam3*, and in both the transposons, a single precursor transcript is predicted to encode two polypeptides.

MECHANISM OF TRANSPOSITION

The mechanism of transposition is not well understood but models for the mechanism have to include the following:

1. Association of the two ends of transposable element
2. Endonuclease activity to cut close to the end of the integrated element or at the target site
3. Ligation of chromosome ends

The following are some of the features common to the *Ac* and *Spm* elements that provide some experimental evidences for these processes.

1. An element-encoded protein (transposase) that binds to *cis*-regulatory regions
2. Asymmetry of *cis*-regulatory regions
3. Terminal IRs, as possible recognition sites for an endonuclease
4. Duplication of bases at the target site

Transposition in a Heterologous Species

Transposable elements have been cloned and introduced into a number of different (heterologous) plant species. A method for measuring transposition in a heterologous species is shown in Figure 4.4.

In the figure, part (1) shows three constructs that contain both the gene for kanamycin resistance (*NptII*) and the gene for hygromycin resistance (*Hpt*). Plant cells will not grow in the presence of either of these two antibiotics. However, introduction of these constructs into the nuclear genome of plants will allow the resulting transformed cells to grow in the media containing these antibiotics. Two of the constructs shown in Figure 4.4 part (1) have a transposable element in the *NptII* gene and these insertion mutations prevent these constructs conferring kanamycin resistance to the recipient plant cells. It is possible to take cells from the leaves of a tobacco plant (*Nicotiana tabacum*) and to use an enzyme cocktail to remove the cell walls from them.

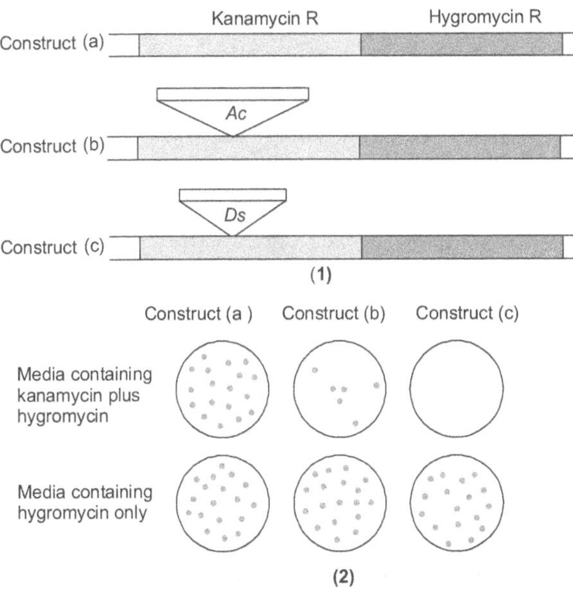

Figure 4.4 Experiment to demonstrate the transposition of *Ac* in a heterologous species. (1) T-DNA constructs containing (a) the genes for hygromycin and kanamycin resistance (b) the gene for hygromycin resistance plus the insertion of autonomous transposable element *Ac* within the kanamycin-resistance gene (c) the gene for hygromycin resistance plus the insertion of the non-autonomous transposable element *Ds* within the kanamycin-resistance gene. (2) Growth of protoplasts containing the constructs (a), (b) or (c) on media containing either hygromycin alone or kanamycin plus hygromycin.

The resulting cells become spherical and are known as protoplasts. The protoplasts can be plated out onto agar plates (Petri dishes) containing a suitable medium, where they reform cell walls and divide to form small colonies of cells.

Part (2) of the figure shows the results of an experiment where the three constructs were introduced separately into three aliquots of a tobacco protoplast suspension. The resulting transformed cells were then plated out onto agar plates containing either kanamycin and hygromycin or hygromycin alone. It also shows that protoplasts containing construct (a) can grow in the presence of both antibiotics. Protoplasts containing construct (c) can grow only on the plate containing hygromycin alone. Because of the presence of the non-autonomous *Ds* transposons in *NptII,* this gene is not expressed and the cells are not resistant to kanamycin.

T-DNA constructs	Kanamycin + hygromycin	Hygromycin only
Construct (a)	Many cell colonies	Many
Construct (b)	Few revertants	Many
Construct (c)	No cell colonies	Many

The result of plating protoplasts containing construct (b) onto plates containing both antibiotics is different from the other two tests, with a small number of cell colonies being formed. This result demonstrates the ability of the *Ac* transposable element to excise from the *Npt*II gene, thereby restoring gene activity and kanamycin resistance of the cells and allowing the formation of a small number of colonies.

TRANSPOSON TAGGING

A number of transposable elements have been cloned and have been used to isolate mutant genes from genomic libraries produced from plants containing a homologous transposon insertional mutation that is containing the same transposon. This strategy used the transposon DNA as a probe for the mutant allele, followed by identification of flanking sequences, which represent the gene, and subsequent use of this DNA to isolate the gene from a wild-type plant genomic library.

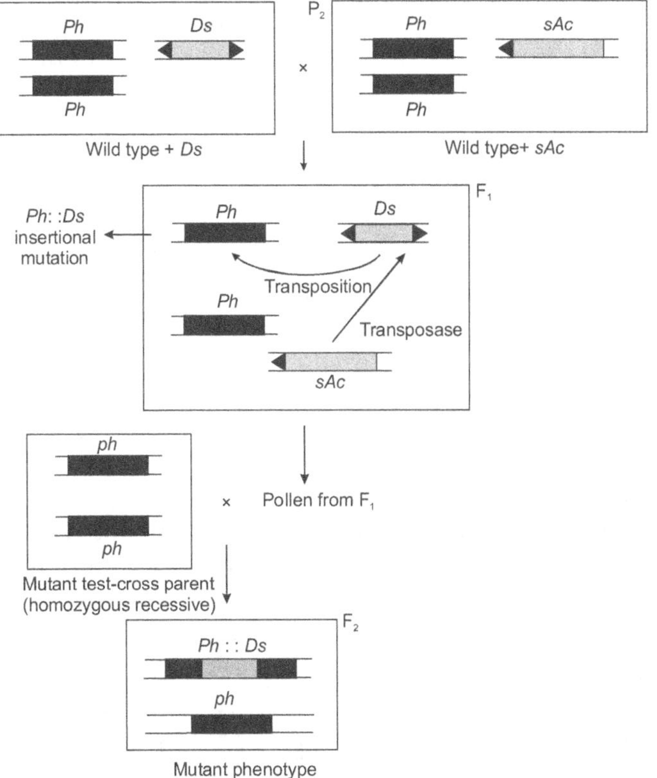

P₁ and P₂—Parent plants containing wild-type dominant *Ph* alleles and non-mobile transposable elements (*Ds* or *sAc*); F₁—First filial generation containing *Ds* (with two inverted repeats) and *sAc* (producing transposase).

Figure 4.5   Transposon tagging. Schematic diagram of a strategy to use the cloned non-autonomous transposable element *Ds* to identify clones of the gene *Ph* by marking the gene using insertional mutagenesis.

Since transposable elements can be introduced and mobilized in a plant, screening procedures can be devised to isolate insertional mutations in genes of known function. This strategy for cloning genes is known as **transposon tagging** and it is dependent upon the gene in question having a known and easily identified mutant phenotype. The technique has been used successfully to clone the *DRL*1 locus of *Arabidopsis*, mutations of which have abnormal leaves and roots, and no inflorescence. Other examples are the *knotted* and *opaque-2* genes of *Zea mays* that are involved in leaf development and regulation of other genes respectively. An outline of transposon tagging is shown in Figure 4.5.

Transposition of *Ds* in these plants may give rise to *Ph* insertional mutations which can be identified by crossing with a homozygous mutant test-cross plant.

A *Ds* element and a stabilized *Ac* (*sAc*) element have been introduced separately into the genome of separate wild-type (normal) plants. These plants are crossed and some of the resulting F_1 progeny will have inherited both the *sAc* and *Ds* elements. Since *sAc* produces transposase, the *Ds* element is mobilized in these F_1 plants and may by chance be inserted into the gene of interest (*Ph*) to produce an insertional mutation, *Ph::Ds*. If this mutation produces a recessive allele that gives rise to a change of phenotype in homozygous individuals, plants containing the *Ph::Ds* mutation can be identified by crossing the F1 plants with homozygous stable recessive plants (*ph/ph*).

Most of the progeny of this cross will be herterozygous, *Ph/ph*, with a normal phenotype, but a few plants will inherit the *Ph::Ds* allele from the F_1 parent, and since the other parent donates a stable recessive allele (*ph*), these plants (*Ph::Ds/ph*) will have a mutant phenotype.

It has been found that the efficiency of this technique can be improved by selecting plants that contain a *Ds* linked to the gene of interest. This is because transposition occurs more frequently into adjacent or flanking regions of DNA than into other parts of the genome. In *Arabidopsis*, the rate of *Ds* transposition is not very high and it has been possible to increase this by changing the transposase promoter within the *Ac* element. Thus substituting the cauliflower mosaic virus (CaMV) 35S constitutive promoter and removing all but 65 bp of the *Ac* transposase, improved the frequency of *Ds* transposition in *Arabidopsis*.

REVIEW QUESTIONS

1.   Give an account of *Ac* and *Ds* elements and their applications.

2.   Give an account of *Tam* transposable elements in *Antirrhinum*.

PART II

PLANT GROWTH AND REGULATION OF GENE EXPRESSION

Plant growth is affected by internal and external factors. The internal controls are all the products of the genetic instructions carried in the plant. These influence the extent and timing of growth and are mediated by signals of various types transmitted within the cell, between cells, or all around the plant. Intercellular communication in plants may take place via hormones (or chemical messengers) or by other forms of communication not well understood. There are several hormones (or groups of hormones), each of which may be produced in a different location, that have a different target tissue and act in a different manner. Hormones namely, auxins, cytokinins, gibberellins, abscisic acid and ethylene play a major role in plants and are hence discussed here.

Molecular biological techniques have considerably increased our understanding of mechanisms underlying seed germination, development, flower initiation and senescence. These processes are critically important in managing the yield of agricultural crops. Hence the molecular and genetic control of plant morphogenesis is delineated.

The ability to regulate both the level and the duration of gene expression permits the study of proteins whose constitutive expression may not be tolerated by the host cell. Most of the available inducible systems involve regulation by antibiotics, hormones, or heat-shock.

5

GROWTH HORMONES

INTRODUCTION

The **phenotype** (physical trait) of an organism is the result of complex interactions between its **genotype** (genetic instructions) and the external **environment.** The growth and differentiation of cells in different parts of the plant are coordinated in response to the genetic code and to the inputs from the environment. The link between the input and the response can be better understood by learning about plant hormones.

Plant hormones are small organic compounds produced by the plant, which influence its physiological responses to the environmental stimuli and act as signalling molecules that have significant effects on growth and development, although they are present at very low concentrations (generally less than 10^{-7} M). They are not directly involved in metabolic or developmental processes, but are indirectly involved by modifying these processes. They are used extensively in agriculture, horticulture, and biotechnology to modify plant growth and development. They also regulate or influence a range of cellular and physiological processes, including the following.

- Cell division

- ✘ Cell enlargement
- ✘ Cell differentiation
- ✘ Flowering
- ✘ Fruit ripening
- ✘ Movement (tropisms)
- ✘ Seed dormancy
- ✘ Seed germination
- ✘ Senescence
- ✘ Leaf abscission and
- ✘ Stomatal conductance

Plant hormones differ from animal hormones in the following aspects.

Plant hormones	Animal hormones
Not made in specialized tissues	Made in specialized tissues
No definite target areas	Definite target areas
No specific effect	Have specific effects

Many plant biologists use the term "plant growth regulators" instead of "plant hormones" to indicate this fact.

Other signalling molecules that are involved in resistance to pathogens and herbivores are jasmonic acid, salicylic acid and the protein systemin. Other "hormonelike" substances produced by plants include polyamines, jasmonates, salicylic acid, brassinosteroids, florigens, phytochromes (photoreceptors) and nitric oxide.

Five major classes of plant hormones are recognized:

1. Auxin—the growth hormone responsible for cell elongation

2. Gibberellin—induces cell elongation and cell division that is translated into growth

3. Cytokinin—brings about cell division and causes inhibition of senescence

4. Abscisic acid—induces abscission of leaves and fruits and is involved in induction of dormancy in buds and seeds

5. Ethylene gas—promotes senescence, epinasty, and fruit ripening

AUXIN—THE GROWTH HORMONE

It was the first plant hormone that was discovered and identified chemically, and was isolated from human urine. However, the functions of auxins in animals are not known.

Pioneering Experiments

In the 19th century, Charles Darwin and his son Franscis studied the phenomenon of plant growth. They conducted their experiment on many grasses in which the youngest (four-day-old) seeds are sheathed in a protective organ called coleoptile (Figure 5.1a). The coleoptiles are very sensitive to light. Hence, when they are illuminated with a short pulse of light, they bend (grow) towards the source of light, within an hour. The curvature formed is shown in Figure 5.1b. This phenomenon of bending of plants towards light is called phototropism.

When the tip of the coleoptile was excised, there was no curvature (Figure 5.1c). Further when they covered the tip of the coleoptile with a foil, no curvature was observed, i.e., it failed to bend towards light (Figure 5.1d). Thus, they found that it was the tip of the coleoptile that perceived the light. The region of coleoptile below the tip that bends in the presence of light is called the **growth zone.** They concluded that the growth of coleoptiles towards light is controlled or influenced by some compound present in the apical region.

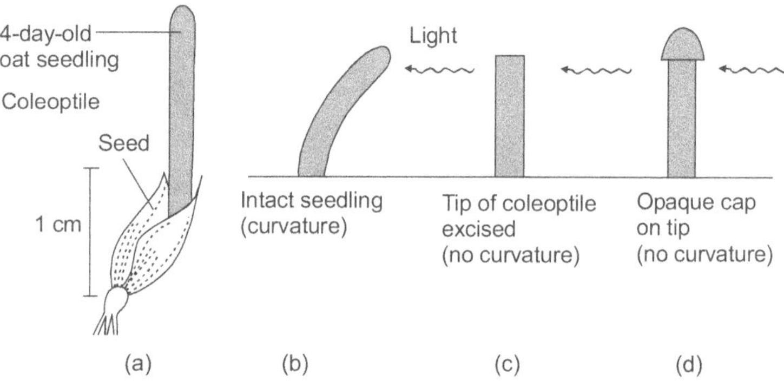

Figure 5.1   Experiment by Charles Darwin and Franscis on growth-inducing compound

In 1913, Peter Boysen and Jensen cut the tips of coleoptiles and placed an intervening water-impermeable barrier in the form of mica and (Figure 5.2a) water-permeable barrier in the form of gelatin (Figure 5.2b), between the cut tip and the cut coleoptile.

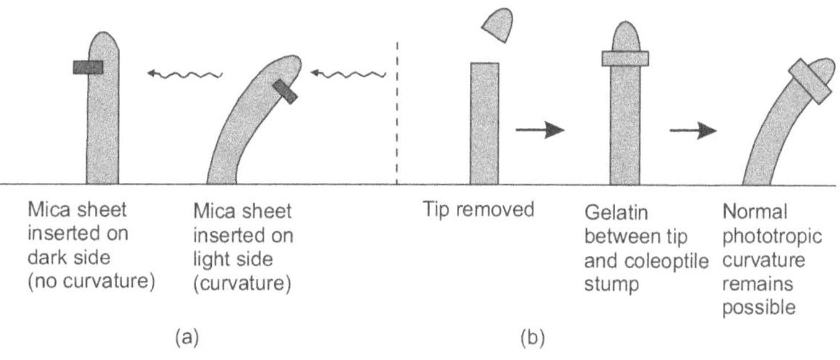

Figure 5.2   Boysen–Jensen experiment

This was done in order to identify the growth-inducing compound. Boysen and Jensen discovered that the growth stimulus passes through materials such as gelatin but not through water-impermeable barriers such as mica. The "influence" passed through the gelatin or agar and re-established the phototropic response. Thus, they concluded that the compound was a water-soluble substance, which produced some sort of signal that travelled to the growth zone to cause bending.

Frists Went, who worked on oat (*Avena sativa*) coleoptile in 1926, observed that if the tip of the coleoptile was removed, the coleoptile stopped growing further. In order to identify the growth-inducing compound present in the tip of the coleoptile, he cut a few coleoptile tips, placed them on a gelatin sheet allowing the chemical to diffuse into it (capillary action) for sometime, and cut the gelatin sheet into small blocks (Figure 5.3). When he placed one of these blocks asymmetrically on top of a **decapitated** coleoptile, it bent away from the side containing the block. This confirmed the presence of a substance that promotes elongation and bending of the coleoptile. This substance was named **auxin** (in Greek, *auxein* means "to grow"). Thus, he showed that some substance promoted the elongation and bending of the coleoptile.

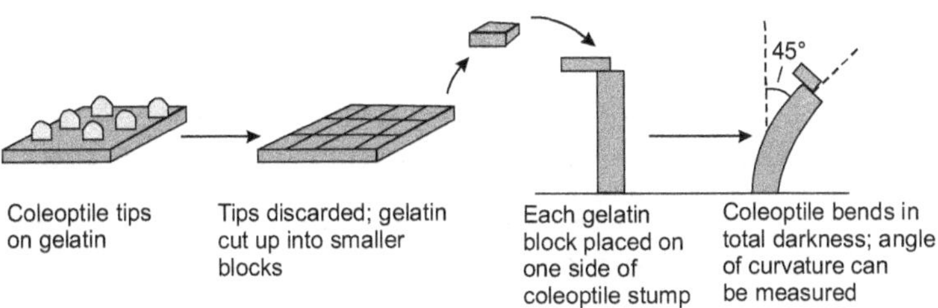

Figure 5.3 Experiment by F. Went

He also devised a coleoptile-bending assay for quantitative auxin analysis.

Biosynthesis and Metabolism

The principal auxin in higher plants is indole 3-acetic acid. Several other auxins in higher plants were discovered later. In the mid-1930s, it was determined that the principal auxin is **indole 3-acetic acid (IAA).** IAA is derived from the amino acid **tryptophan.**

Tryptophan

(The precursor)

Indole 3-acetic acid (IAA)

Biosynthesis of Auxin

IAA is synthesized in meristems, young leaves, hydathodes and developing fruits and seeds. Its most active areas of synthesis are in young leaves, fruits, flowers, shoot tips, embryos, and pollen. Free IAA is the biologically active form of the hormone. However, a vast majority of auxin in plants is found in a covalently bound state. These conjugated, or "bound" auxins are considered hormonally inactive. Based on their mass, IAA conjugates have been categorized into two types.

1. Low-molecular-weight conjugated auxins (e.g. esters of IAA with glucose, myo-inositol, amide conjugates)

2. High-molecular-weight IAA conjugates (e.g. IAA glucan, IAA glycoproteins)

IAA biosynthesis (Figure 5.4) is associated with rapidly dividing and rapidly growing tissues, especially shoots. IAA is structurally related to the amino acid tryptophan. Recent genetic studies in plants also indicate the existence of one or more tryptophan-independent pathways for the synthesis of IAA. The tryptophan-dependent pathways are shown in Figure 5.4.

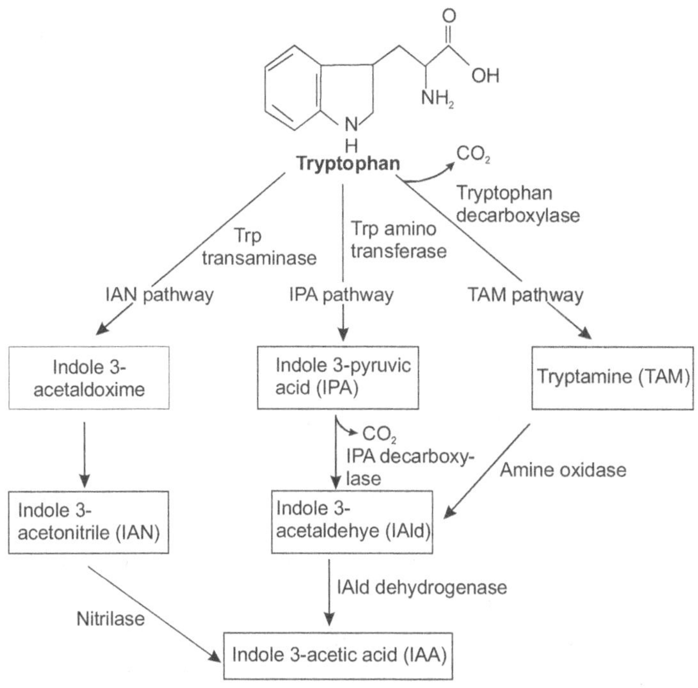

IAld—Indole 3-acetaldehyde dehydrogenase; Trp—Tryptophan

Figure 5.4   Tryptophan-dependent pathways of IAA biosynthesis in plants

1. *The indole pyruvic acid (IPA) pathway*   Tryptophan is first deaminated to form IPA, followed by a decarboxylation reaction to form indole 3-acetaldehyde (IAld), which is then oxidized to IAA by IAld dehydrogenase.

2. *The indole 3-acetonitrile (IAN) pathway*   Here tryptophan is first converted to indole 3-acetaldoxime and then to indole 3-acetonitrile (IAN). Finally it is converted to IAA by nitrilase.

3. *The tryptamine (TAM) pathway*   Tryptophan is first decarboxylated and then deaminated (reverse of IPA pathway) to form IAA.

The existence of these multiple pathways for IAA biosynthesis makes it nearly impossible for plants to run out of auxin, and this highlights its important role in plant development.

Degradation of IAA   IAA is degraded to oxindole 3-acetic acid (OxIAA). *In vitro*, IAA can be oxidized non-enzymatically when exposed to high intensity of light.

Transport

From the experiments of Went, it was established that IAA moves mainly from the apex to the basal end (basipetally), in excised coleoptile sections. This type of unidirectional transport is termed as **polar transport**. Auxin is the only plant growth hormone that is known to be transported thus. During its transport from the shoot tip to the root tip, a gradient is established and hence auxin participates in various developmental processes including stem elongation, apical dominance, wound healing, and leaf senescence.

Tissues differ in the degree of polarity of IAA transport. In coleoptiles, vegetative stems, and leaf petioles, basipetal transport is predominant. Polar transport is not affected by gravity. Polar transport proceeds in a cell-to-cell fashion rather than via the plasmodesmata, i.e., auxin exits the cells (auxin efflux) through the plasma membrane, diffuses across the compound middle lamella and enters the cells (auxin influx) below through its plasma membrane. The overall process requires energy.

Very recently, it was found that auxin transport occurs also in the phloem. This is the principal route where auxin is transported **acropetally** (i.e., towards the tip from root cap) in the root. In the phloem, auxin transport is non-polar and most of the IAA synthesized in the mature leaves is transported to the rest of the plant via the phloem. Auxin along with the other compounds of phloem sap moves from these leaves up or down the plant and this is a passive transport not requiring the energy directly.

Physiological Effects

1. *Auxin promotes cell elongation (growth) in stems and coleoptiles while inhibiting growth in roots.* Auxin is synthesized in the shoot apex and transported basipetally to the tissues below. The steady supply of auxin stimulates the rate of elongation of cells. The optimal concentrations of auxin are in the range of 10^{-6} to 10^{-5} M. Low concentrations (10^{-10} to 10^{-9} M) of auxin promote the growth of intact roots, whereas high concentrations (10^{-6} M) inhibit root growth.

2. Lateral redistribution of auxin is responsible for phototropism and gravitropism in plants.

3. *Auxin rapidly increases the extensibility of the cell wall.* Plant cells expand in 3 steps.

    i. Osmotic uptake of water across the plasma membrane is driven by water potential.

    ii. Turgor pressure builds up because of the rigidity of the cell wall.

    iii. Wall loosening occurs, allowing the cell to expand in response to turgor pressure.

4. *Auxin regulates apical dominance.* In most of the higher plants, the growing apical bud inhibits the growth of lateral (axillary) buds, a phenomenon called **apical dominance**. Removal of the shoot apex (decapitation) usually results in the growth of one or more of the lateral buds. The hypothesis is that the

outgrowth of axillary bud is inhibited by auxin that is transported basipetally from the apical bud.

5. *Auxin promotes the formation of lateral and adventitious roots.* At high concentrations, auxin initiates the formation of lateral and adventitious roots but not the primary root ($>10^{-8}$ M). Lateral roots are commonly found above the elongation and root hair zone and originate from small groups of cells in the pericycle. Auxin stimulates these pericycle cells to divide. The dividing cells gradually form a root apex and the lateral root grows through the root cortex and epidermis.

6. *Auxin delays the onset of leaf abscission.* The shedding of leaves, flowers and fruits from a living plant is known as **abscission**. These parts abscise in a region called the **abscission zone** which is located near the base of the petiole of leaves. During leaf senescence, a distinct layer of cells in **the abscission layer** (within the abscission zone) is digested causing the leaf base to become soft and weak. The weak base causes the leaf to abscise. Auxin levels are high in young leaves, progressively decrease in maturing leaves and are low in senescing leaves. This suggests that abscission is triggered when auxin is no longer being produced.

Therefore, auxins have the ability to promote the following.

1. cell elongation in coleoptile and stem pieces.

2. cell division in callus cultures (in the presence of cytokinins).

3. formation of adventitious roots on detached leaves and stems.

Molecular Mode of Action

Growth rate studies on both the excised stem and coleoptiles show that on addition of auxin, a rapid stimulation of growth occurs after a lag period of only 10 to 12 minutes. This stimulation requires energy and proteins with high turnover rates. The lag time of 10 minutes is the time required by the biochemical machinery of the cell to bring about the increase in the growth rate.

The basis for growth is the auxin-induced proton extrusion that causes acidification of the cell wall leading to cell extension. According to the acid growth hypothesis, hydrogen ions act as the intermediate between auxin and cell wall loosening. The plasma membrane H^+ATPase is a primary active transporter that pumps H^+ ions from the cytosol to the outside.

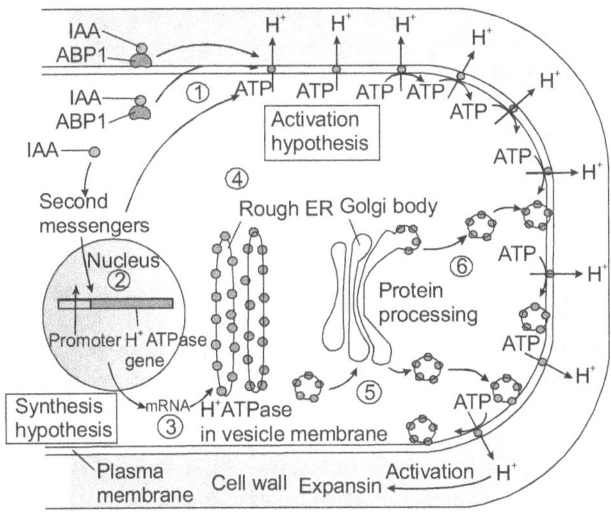

Activation hypothesis—binding of IAA to ABP1 (auxin binding protein) causes activation of H^+ATPase; Synthesis hypothesis—IAA increases the expression of the H^+ATPase gene leading to increase in number of H^+ATPase transporters in the plasma membrane

Figure 5.5 Mechanism of action of auxin

Auxin possibly increases the rate of proton extrusion either by activating the pre-existing membrane H^+ATPases or by increasing the synthesis of new H⁺ATPases on the plasma membrane. Some auxin-binding proteins (ABPs) are thought to be involved in the activation model (Figure 5.5).

A wall-loosening protein called expansin has been identified. At acidic pH value, this protein weakens the hydrogen bonds between the polysaccharide components of the wall leading to loosening of the cell wall.

Commercial Uses

Auxins have been used commercially in agriculture and horticulture for more than 50 years. The commercial uses include

    i. prevention of falling of fruits and leaves,

    ii. promotion of flowering in pineapple,

    iii. induction of parthenocarpic (seedless) fruits and

    iv. initiation of adventitious roots in commercial plant cuttings.

GIBBERELLINS—THE REGULATORS OF PLANT HEIGHT

In the 1950s, a second group of hormones was characterized and was named the gibberellins (GAs). The gibberellins are a large group of related compounds (more than 125 are known). They are most often associated with the promotion of stem growth, and the application of gibberellins to intact plants can induce heightening of the plant.

Identification of Gibberellins

In the 1930s, Japanese scientists obtained two growth-active compounds from the fungus, *Gibberella fujikuroi*, which they termed **gibberellin A** and **B**. Simultaneously, Takahashi and Tamura isolated three gibberellins from the original gibberellin A and named them gibberellin A_1, gibberellin A_2 and gibberellin A_3. Later, the US Department of Agriculture named the growth-active compound as **gibberellic acid**. Gibberellin A_3 and gibberellic acid proved to be identical.

Gibberellic acid (GA_3)

Gibberellin A_1 (GA_1)

Biosynthesis and Metabolism

Gibberellins constitute a large family of diterpene acids that are synthesized by a branch of the terpenoid pathway. Plants contain a large array of gibberellins, many of which are biologically inactive. GC–MS (Gas chromatography combined with mass spectrometry) is the most widely used method for gibberellin analysis because active gibberellins are present in very low quantities in plants.

Gibberellins are tetracyclic diterpenoids made up of four isoprenoid units. Terpenoids are compounds made up of isoprene (five-carbon) building blocks. They are synthesized via the terpenoid pathway in three stages (Figure 5.6).

1. The first stage involves the formation of precursor compounds for the synthesis of gibberellins, and it occurs in the plastid.

2. The second stage involves oxidation reactions that produce GA_{12}, the first gibberellin in the pathway in all plants and the precursor of all gibberellins, and it occurs in the ER.

3. The third stage involves the production of GA_{20} through hydroxylation and oxidation reactions, and it occurs in

the cytosol. This GA_{20} is converted to the biologically active form GA_1 by hydroxylation of carbon 3.

GA_1 is the biologically active gibberellin controlling stem growth. Similar to auxins, all gibberellins are biosynthesized in apical tissues.

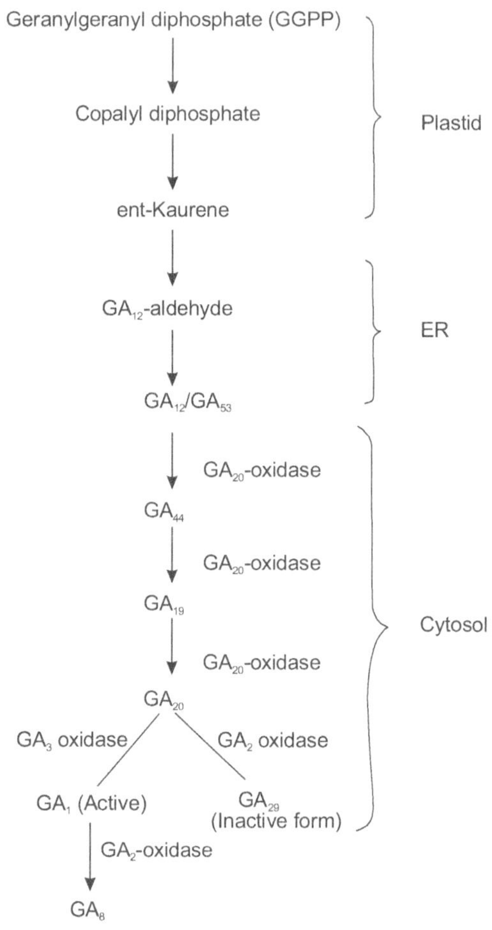

Figure 5.6 Biosynthesis of gibberellins

Gibberellins can regulate their own metabolism. Endogenous gibberellins regulate their own metabolism by either switching on or inhibiting the transcription of the genes encoding the enzymes needed for gibberellin biosynthesis and degradation, i.e., feedback regulation. Thus, the level of active gibberellins is kept within a narrow range. External application of gibberellin causes a down-regulation in the transcription of the biosynthetic genes encoding GA_{20} oxidase and GA_3 oxidase and an elevation in the transcription of the degradative gene, GA_2 oxidase.

Synthesis of gibberellins is promoted by auxin. The presence of auxin has been shown to promote the transcription of GA_3 oxidase gene and to repress the transcription of GA_2 oxidase gene. Thus, auxins promote the biosynthesis of gibberellins. In the absence of auxin, the reverse occurs.

Dwarfness can now be genetically engineered. Reductions in GA_1 levels have recently been achieved in crops such as sugar beet and wheat, either by the transformation of plants with antisense constructs of the GA_{20} ox or by the GA_3 oxidase genes, which encode the enzymes leading to the synthesis of GA_1.

Physiological Effects

Endogenous gibberellins influence a wide variety of developmental processes both in the vegetative and reproductive developments. In the case of vegetative development, gibberellins help in stem elongation. In addition to stem elongation, gibberellins control various aspects of seed germination, including the loss of dormancy and the mobilization of endosperm reserves. In the case of reproductive development, gibberellins help in floral initiation, sex determination and fruit set. The following are the major important roles of gibberellins in the growth of a plant.

1. *Gibberellins stimulate stem growth.* Externally applied gibberellins promote internodal elongation. They also cause a decrease in stem thickness and leaf size and give a pale green colour to the leaves, but it seem to have very little effect on root growth.

2. *Gibberellins regulate the transition from juvenile to adult phases.* Many juvenile conifers can be induced to enter the reproductive phase by the application of gibberellins.

3. *Gibberellins influence floral initiation and sex determination.* In maize, the staminate flowers (male) are restricted to the tassel, and the pistillate flowers (female) are contained in the ear. Application of exogenous gibberellic acid to the tassels induces pistillate formation.

4. *Gibberellins promote fruit set.* Application of gibberellins can cause **fruit set** (the initiation of fruit growth following pollination) and growth of some fruits such as apple.

5. *Gibberellins promote seed germination.* Seed germination requires gibberellins for the mobilization of stored food reserves of the endosperm layer by weakening the cell walls. Some seeds require light or cold to induce germination. In such seeds, this dormancy can be overcome by the application of gibberellin.

Gibberellins also stimulate the production of α-amylase by the aleurone layer of germinating cereal grains. This aspect of gibberellin action has led to its use in the brewing industry in the production of malt.

Commercial Uses

GA_3 is used mainly in orchards, brewing industries and sugar industries. It is either sprayed or used for dipping. In some crops, a reduction in height is desirable and this can be accomplished by the use of gibberellin synthesis inhibitors.

CYTOKININS—THE REGULATORS OF CELL DIVISION

Cytokinins were identified as factors that stimulate the plant cells to divide, i.e., they make the cells undergo cytokinesis.

Discovery and Identification

Letham (1974) discovered cytokinins in coconut milk. This cytokinin was named **zeatin**. However, before this discovery, the first report of the synthetic analogue of cytokinin called **kinetin** came from the lab of Folke Skoog *et al.* in 1950. They found that the autoclaved Herring sperm DNA had a powerful cell-division-promoting effect. After much work, a small molecule was identified from the

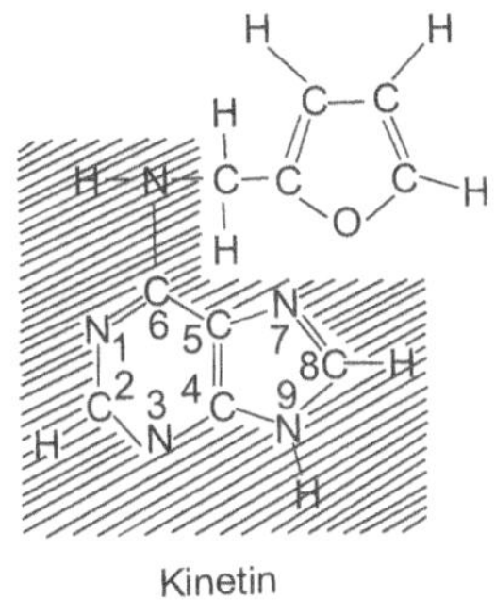

Kinetin

autoclaved DNA and was named kinetin. This was shown to be an adenine or aminopurine derivative, 6-furfuryl aminopurine.

Kinetin is not a naturally occurring plant growth regulator. It is a by-product of the heat-induced degradation of DNA, in which the deoxyribose sugar of adenosine is converted to a furfuryl ring and is shifted from the 9th position to the 6th position on the adenine ring. Benzyl aminopurine (BAP) is an example of a synthetic N^6-substituted aminopurine cytokinin.

Extracts of the immature endosperm of corn (*Zea mays*) were found to contain a substance that had the same biological effect as kinetin. Both the auxins and cytokinins participate in the regulation of the cell cycle and they do so by controlling the activity of cyclin-dependent kinases. Cyclin-dependent protein kinases (CDKs) together with the cyclins (their regulatory subunits) are the enzymes that regulate the eukaryotic cell cycle. The expression of the gene that encodes CDC2 (the major CDK, responsible for cell division cycle) is regulated by auxin.

In the presence of an auxin, kinetin stimulates any tissue to proliferate in culture. If kinetin alone is added to the medium without auxin, it does not induce cell division. In higher plants, zeatin occurs in both *cis* and *trans* configurations, and these forms can be interconverted by the enzyme, **zeatin isomerase**. Both forms play important roles in cell division.

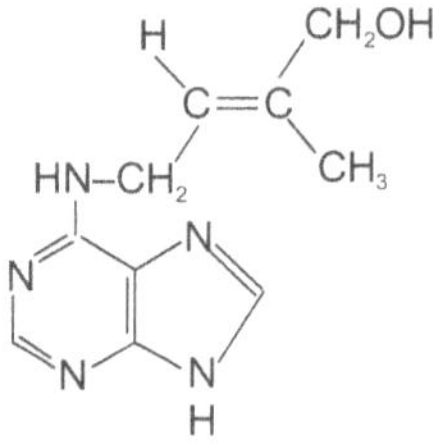

trans-zeatin 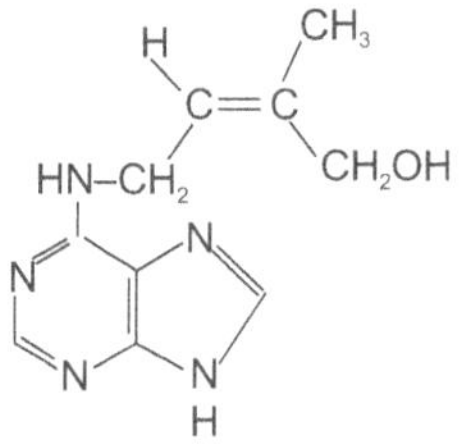   *cis*-zeatin

6-(4-Hydroxy 3-methylbut-2-enylamino) purine

Cytokinins occur in both free and bound forms. Some plant pathogenic bacteria secrete free cytokinins. A typical example is *Agrobacterium tumefaciens* that forms crown galls. *Corynebacterium fascians* also produces cytokinins that cause growth abnormality called **witches broom** in balsam fir.

Biosynthesis

Cytokinins have side chains synthesized from isoprene derivatives. Isoprene units are formed from the precursors mevalonate or pyruvate and 3-phosphoglycerate. These precursors form dimethyl allyl diphosphate, an isomer of isoprene.

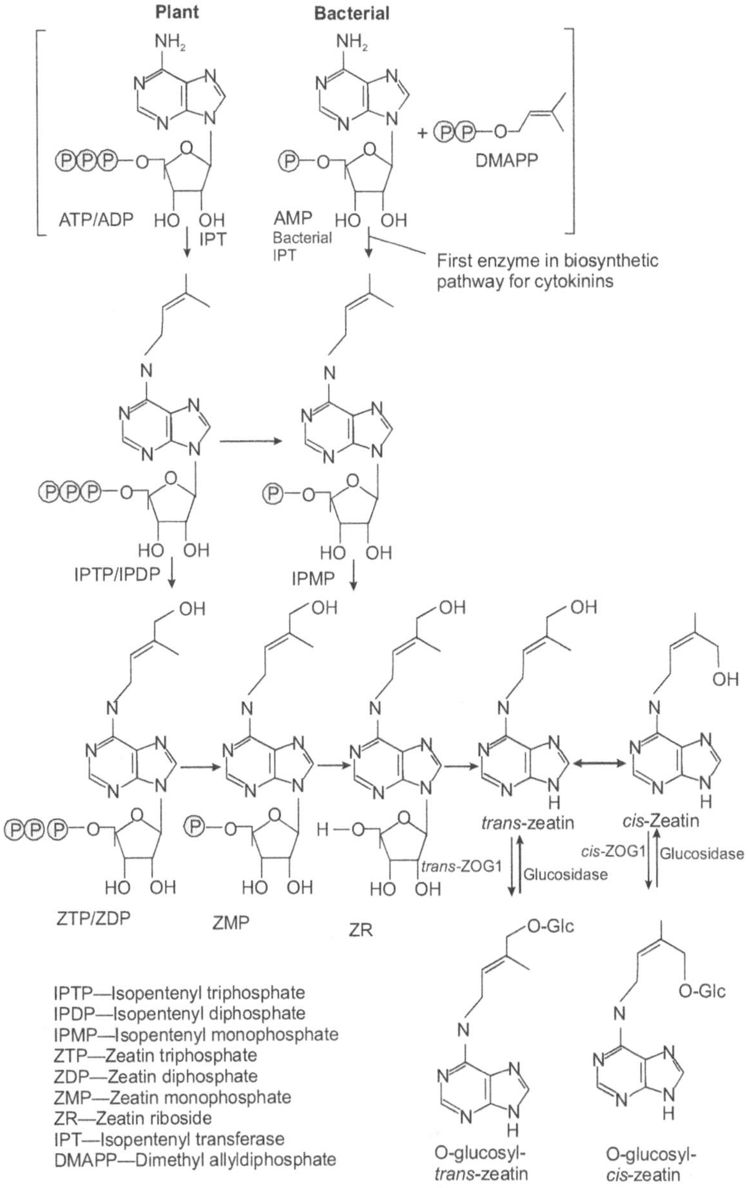

Figure 5.7   Biosynthetic pathway for cytokinin

The first committed step in cytokinin biosynthesis is the addition of the isopentenyl side chain from dimethylallyl diphosphate (DMAPP) to an adenosine moiety. The plant and bacterial isopentenyl transferase (IPT) enzymes differ in the adenosine substrate used. The plant enzyme appears to utilize both adenosine diphosphate and adenosine triphosphate (ADP and ATP) and the bacterial enzyme utilizes adenosine monophosphate (AMP). The products of these reactions (IPMP, IPDP or IPTP) are converted to zeatin by a hydroxylase. The various phosphorylated forms can be interconverted and free *trans*-zeatin can be formed from the riboside by enzymes of general purine metabolism. *trans*-Zeatin can be metabolized in various ways as shown in Figure 5.7.

Physiological Effects

Transport of cytokinin from the root to the shoot via xylem   Root apical meristems are the major sites of synthesis of the free cytokinins in plants. The cytokinins synthesized in roots move through the xylem into the shoot, along with the water and minerals taken up by the roots. Xylem exudate thus helps in cytokinin movement in plants.

Cytokinins are rapidly converted to their respective nucleoside and nucleotide forms. Such interconversions involve enzymes common to purine metabolism. Many plant tissues contain the enzyme **cytokinin oxidase**, which cleaves the side chain from zeatins (both *cis* and *trans*).

Delay of leaf senescence   The abscission zone at the base of a leaf petiole is the region where mature parenchyma cells begin to divide after a period of mitotic inactivity, forming a layer of cells with relatively weak cell walls where abscission can occur.

Leaves detached from the plant slowly lose chlorophyll, RNA, lipids and proteins, although they are kept moist and provided with minerals. This programmed aging process leading to death is called senescence. Leaf senescence is more rapid in dark than in light.

Gan and Amasino (1995) transformed tobacco with *ipt* gene which encodes the first enzyme in the biosynthetic pathway of cytokinins. The gene was placed under the control of a leaf senescence-specific promoter (*Arabidopsis* cysteine protease promoter) and hence the transgenic plants had wild-type levels of cytokinins and developed normally, up to the onset of leaf senescence. As the leaves aged, the cysteine protease promoter was activated, triggering the expression of the *ipt* gene within leaf cells. The resulting elevated levels of cytokinin blocked senescence.

Cytokinins have an effect on the following processes:

- leaf senescence
- nutrient mobilization
- apical dominance
- formation of active shoot apical meristems
- floral development
- breaking of bud dormancy
- seed germination

Molecular Mode of Action

Cytokinin response is mediated by a two-component regulatory system (Figure 5.8) composed of two functional elements. **i. a sensor histidine kinase** to which a signal binds, and **ii. a downstream response regulator** whose activity is regulated by histidine kinase. The sensor histidine kinase is a membrane-bound protein that contains 2 distinct domains called the input and the transmitter domains. The response regulator contains two domains the receiver domain and the output domain.

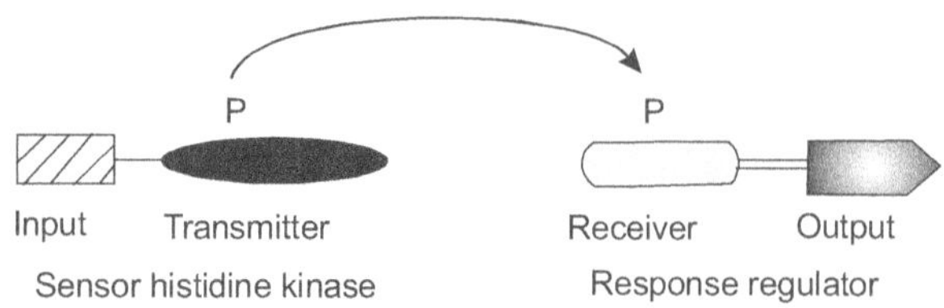

Figure 5.8   Two-component regulatory system

Upon detection of a signal, the sensor histidine kinase (a dimer) transphosphorylates a conserved histidine residue. This phosphate is then transferred to a conserved aspartate residue in the receiver domain of the response regulators. These response regulators are mostly the transcription factors that help in the activation of transcription of genes associated with cytokinin biosynthesis.

Application of Cytokinins

Application of cytokinin to plants leads to the triggering of other genes that encode

1. nitrate reductase
2. light-regulated proteins
3. pathogenesis-related proteins
4. rRNAs
5. peroxidases

ABSCISIC ACID (ABA)

In 1963, a substance that promoted the abscission of cotton fruits was purified and crystallized. This was termed **abscisin II**. At the same time, a substance that promoted bud dormancy was purified from *Sycamore* leaves and was called **dormin**. Later, it was found to be abscisin II and hence was renamed as abscisic acid (ABA).

It is a ubiquitous plant hormone occurring in every living tissue from the root cap to the apical bud (Milborrow, 1984).

Chemical Structure

ABA is a fifteen-carbon terpenoid compound derived from the terminal portion of carotenoids (Figure 5.9). The fifteen carbon atoms of ABA form an aliphatic ring with one double bond, three methyl groups and an unsaturated chain that has a terminal carboxyl group. The orientation of the carboxyl group at carbon 2 determines the *cis-* and *trans-*isomers of ABA. The naturally occurring ABA is in the *cis-*form.

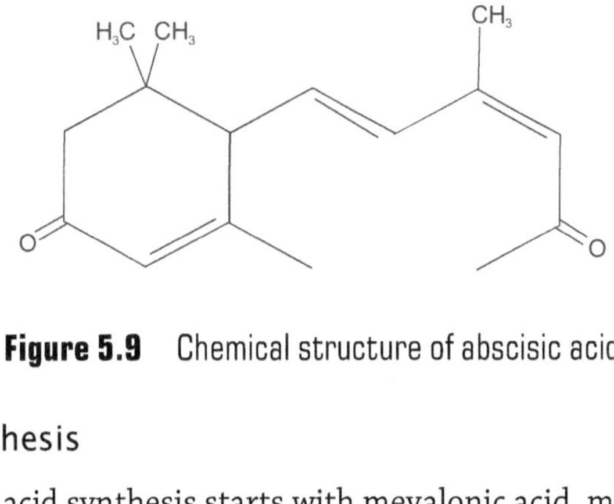

Figure 5.9   Chemical structure of abscisic acid

Biosynthesis

Abscisic acid synthesis starts with mevalonic acid, mentioned earlier in the biosynthesis of cytokinins. Mevalonic acid is converted to farnesyl pyrophosphate, a C_{15} compound, via several intermediates. The subsequent conversion of farnesyl pyrophosphate to zeaxanthin, a C_{40} carotenoid, again involves multiple steps. The transformation of zeaxanthin to all-*trans*-violaxanthin consists of two epoxidations at the double bonds in both cyclohexenyl rings, with antheraxanthin as an intermediate. This reaction is catalysed by zeaxanthin epoxidase (ZEP), the enzyme coded by *aba*1. All-*trans*-violaxanthin undergoes two isomerization reactions to yield 9-*cis*-neoxanthin, which is then cleaved to form a C_{15} compound, xanthoxal. This cleavage is catalysed by the enzyme 9-*cis*-epoxy carotenoid dioxygenase (NCED) that is located on the thylakoid.

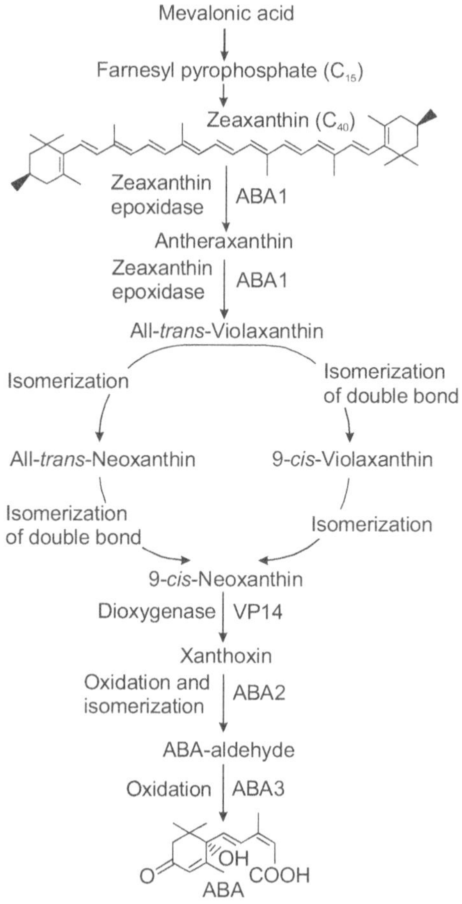

Figure 5.10   Biosynthesis of abscisic acid

The subsequent oxidative cleavage of xanthoxin is thought to be the rate-limiting step in ABA biosynthesis. Xanthoxin is subsequently converted to ABA-aldehyde by oxidation and isomerization. A final oxidation, catalysed by ABA-aldehyde oxidase, produces ABA (Figure 5.10).

Transport

Concentration of ABA can vary in specific tissues especially in the developing seedlings. It increases during water stress, and upon rewatering, there is a decrease in the rate of synthesis. ABA is synthesized in almost all cells that contain plastids and is transported via both the xylem and the phloem. The ABA synthesized in the roots is transported to the shoot via the xylem.

Physiological Roles

1. *ABA promotes desiccation tolerance in the embryo.* It protects the developing seeds from desiccation by synthesizing the proteins involved in desiccation tolerance and is required for the synthesis of storage proteins and seed dormancy. The levels of ABA in seeds peak during embryogenesis. Typically, the ABA content of seeds is very low early in the embryogenesis, reaches a maximum at about the halfway point, and then gradually falls to low levels as the seed reaches maturity. During the mid to late stages of seed development, specific mRNAs encode the so-called late-embryogenesis-abundant (LEA) proteins that help the seeds in desiccation tolerance. The synthesis of these LEA proteins is under the control of ABA.

2. *ABA promotes the accumulation of seed storage protein during embryogenesis.* Storage compounds accumulate in the mid-phase of embryogenesis. ABA-deficient mutants have reduced storage proteins. ABA maintains the mature embryo in a dormant state until the environmental conditions are optimal for growth.

3. *ABA inhibits precocious germination and vivipary.* When ABA is added to a culture medium, it prevents the seeds from germinating. In maize, several viviparous (vp) mutants have

been selected in which the embryos germinate directly on the cob while still attached to the plant. Several of these mutants (vp2, vp5, vp7 and vp14) were ABA-deficient.

4. *Seed dormancy is controlled by the ratio of ABA to GA.* Seeds that are released from the plant in a dormant state are said to exhibit **primary dormancy**. Seeds that are released from the plant in a non-dormant state but which become dormant if the conditions are unfavourable are said to exhibit **secondary dormancy**.

ABA mutants have been useful in finding out the role of ABA in seed dormancy. ABA-deficient (aba) mutants of *Arabidopsis* produced seeds that were non-dormant at maturity.

In many plants, high levels of ABA coincide with a decrease in indole acetic acid (IAA) and gibberellic acid (GA) levels. The following experiment revealed a correlation between ABA and GA. The seeds of a GA-deficient mutant that could not germinate in the absence of exogenous GA were again mutagenized and then grown in a greenhouse. The seeds produced by these mutagenized plants were then screened for revertants, that is, seeds that regained their ability to germinate. When these revertants were analysed, they were found to be the mutants of ABA synthesis. The revertants immediately germinated because dormancy had not been induced and hence they did not require GA to overcome dormancy. Therefore, embryo dormancy is thought to be due to the presence of inhibitors, especially ABA, as well as the absence of growth promoters, such as GA (gibberellic acid). The loss of embryo dormancy is often associated with a sharp drop in the ratio of ABA to GA.

5. *ABA promotes root growth and inhibits shoot growth at low water potentials.* ABA has different effects on the growth of roots and shoots and the effects are strongly dependent on the

water status of the plant. When water supply is ample (high water potential), shoot growth is greater in the wild-type plant (normal endogenous ABA levels) than in the ABA-deficient mutant. Shoot growth is reduced in the ABA-deficient mutant plants. This reduced shoot growth in the ABA-deficient mutants could be due to the water loss from leaves. When the availability of water is limited (i.e., at low water potentials), the reverse occurs. Shoot growth is reduced in wild type plants where as it is greater in the ABA-deficient mutant than in the wild type at low water potential. These observations suggest the following:

1. The endogenous ABA promotes shoot growth in well-watered plants.

2. Endogenous ABA acts as a signal to reduce shoot growth only under water stress conditions.

6. *ABA promotes leaf senescence.* ABA has been shown to be clearly involved in leaf senescence. When ABA promotes senescence, it also indirectly increases the ethylene formation that further helps in senescence.

Cellular and Molecular Modes of Action

ABA is involved in short-term physiological effects (e.g. stomatal closure) as well as long-term developmental processes (e.g. seed maturation). Short-term responses involve some alterations in the fluxes of ions across membranes and in gene regulation. Long-term processes involve major changes in the pattern of gene expression.

Signal transduction pathways amplify the primary signal generated when the hormone (ABA) binds to its receptors. These pathways are required for both the short-term and the long-term effects of ABA.

Short-term effects

1. ABA is perceived both intracellularly and extracellularly. It is widely assumed that there are membrane receptors (proteins) for ABA. This extracellular perception of ABA prevents stomatal opening. Intracellular perception of ABA also induces stomatal closure and inhibits the K^+ current required for opening.

2. ABA signalling involves Ca^{2+}-independent pathways. ABA increases the concentration of cytosolic Ca^{2+} by releasing these ions from central vacuole, leading to stomatal closure. However, it has also been shown to close the guard cells even in the absence of increase in cytosolic calcium. In other words, ABA seems to act via a calcium-independent pathway (Allan *et al.,* 1994).

3. ABA increases cytosolic Ca^{2+}, raises cytosolic pH and depolarizes the membrane. Stomatal closure is driven by a reduction in guard cell turgor pressure caused by a massive efflux of K^+ and anions from the cell. The shrinkage of cells occurs due to water loss and the surface area of the membrane is reduced by nearly 50%. The shrunken parts of the membrane are taken up as small vesicles by endocytosis. Membrane polarization coincides with an increase in cytosolic calcium concentration.

Enzymes involved in ABA response   The signal transduction in guard cells involves the enzymes, protein kinases, phosphatases and ATP. Addition of the protein kinase inhibitors inhibits the ABA-induced stomatal closing. Protein phosphorylation and dephosphorylation play important roles in the ABA signal transduction pathway in guard cells.

There is a direct evidence for an ABA-activated protein kinase (AAPK) in *Vicia faba* guard cells (Mori & Muto, 1997). AAPK activity is required for ABA activation of anion currents and stomatal closing. In addition, two Ca^{2+}-dependent protein kinases as well as MAP kinases, have an effect on ABA regulation of stomatal aperture.

Long-term effects ABA regulation of gene expression is mediated by transcription factors. ABA causes changes in gene expression. It regulates the expression of many genes during seed germination and under stress conditions such as heat shock, low temperature and salt tolerance. The genes include those encoding proteases, chaperonins, and enzymes of sugar metabolism, transcription factors and protein kinases.

A few DNA elements have been identified that are involved in transcriptional repression by ABA. The best characterized of these are the gibberellin response elements (GAREs) that mediate the gibberellin-inducible, ABA-repressible expression of the barley α-amylase gene.

Four transcription factors involved in ABA gene activation in maturing seeds have been identified by genetic means. Mutations in the genes encoding any of these proteins reduce seed ABA response. The maize *vp*1 (viviparous-1) and *Arabidopsis abi*3 (ABA-insensitive 3) genes code for transcription factors. The transcription factors coded by *vp*1, *abi*4 and *abi*5, can either activate or repress the transcription, depending on the target gene.

Other negative regulators of the ABA response Mutants showing enhanced responses to ABA have been isolated. These mutants include *era* (enhanced response to ABA) and *abh* (ABA hypersensitive) lines. These are resistant to wilting and are mildly drought-tolerant.

ETHYLENE—THE GASEOUS HORMONE

Plants produce ethylene both during developmental changes and in reaction to environmental factors. The structure of ethylene is simple and is given by

$$H_2C = CH_2$$
Ethylene

It is involved in a wide range of physiological responses and in some cases it may have the opposite effects in different species. For example, in most dicotyledonous species, ethylene inhibits the stem elongation; however in some aquatic dicotyledons and in rice, it stimulates shoot growth.

Triple Response

Neljubow in 1901 discovered the dramatic effect of ethylene on the growth of etiolated (dark-grown) dicotyledonous seedlings. This is known as the triple response (Figure 5.11) because it causes the following three effects.

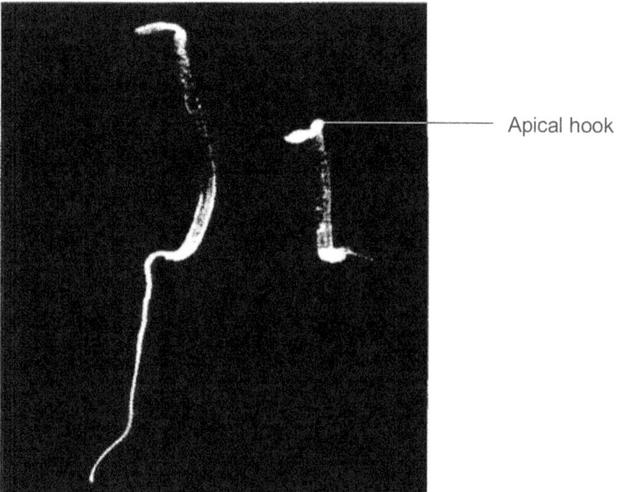

Figure 5.11 The triple response in *Arabidopsis*. Three-day-old etiolated seedlings grown in the presence (right) or absence (left) of 10 ppm ethylene. Note the shortened hypocotyl, reduced root elongation and exaggeration of the curvature of the apical hook that result from the presence of ethylene.

1. Inhibition of the epicotyl (the base of the seedling stem between the shoot apex and cotyledons) or hypocotyl (the region between the cotyledon and the root) elongation and the root elongation.

2. Radial swelling of epicotyl or hypocotyl, and root cells.

3. Exaggeration of the apical hook.

At concentrations above 0.1 mL L^{-1}, ethylene changes the growth pattern of seedlings by reducing the rate of elongation, and by increasing the lateral expansion leading to swelling of the region below the hook.

Etiolated seedlings are usually characterized by the **hook-shaped** terminal portion of the shoot apex. This shape facilitates the movement of the seedling through the soil and protects the tender apical meristem. When it is exposed to light, the hook opens because the elongation rate of the inner side increases. As long as ethylene is produced by the hook tissue in the dark, the elongation of the cells on the inner side is inhibited.

Biosynthesis

Ethylene can be produced by almost all parts of higher plants. The rate of production depends on the type of tissue and the stage of development. In general, meristematic regions and nodal regions are the most active in ethylene biosynthesis. Ethylene production increases during leaf abscission and flower senescence, as well as during fruit ripening. Wounding can also induce ethylene biosynthesis.

Methionine is the precursor of ethylene in higher plants, which initially reacts with ATP to produce S-adenosylmethionine (Ado-Met) in the ethylene biosynthetic pathway (Figure 5.12). ACC synthase is the enzyme that catalyses the conversion of Ado-Met to 1-aminocyclopropane 1-carboxylic acid (ACC). Finally, the ACC is oxidized to ethylene by the enzyme ACC oxidase.

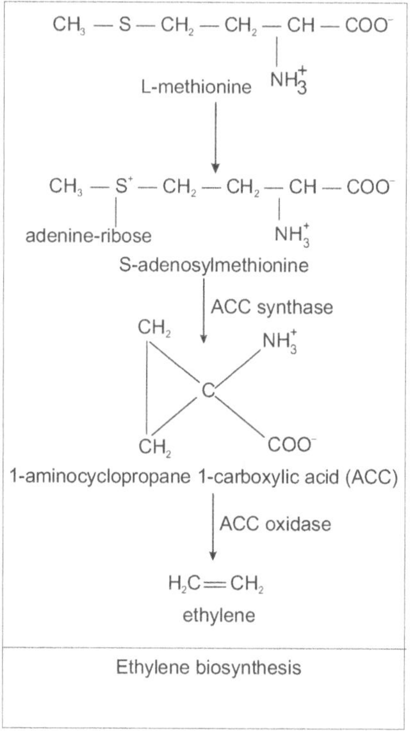

Figure 5.12   Biosynthesis of ethylene

The factors stimulating ethylene biosynthesis are

1. fruit ripening

2. environmental stress

3. auxin

As the fruits mature, the rate of ethylene biosynthesis increases. Fruit maturation or ripening refers to the changes in a fruit that make it ready to eat. Such changes typically include softening due to the enzymatic breakdown of the cell walls, starch hydrolysis, sugar accumulation and the disappearance of organic acids and phenolic compounds. For many years, ethylene has been recognized as the hormone that accelerates the ripening of edible fruits. Addition of ethylene to such fruits hastens the ripening and a dramatic increase in ethylene production is closely associated with the initiation of ripening. Ethylene biosynthesis is increased by stress conditions such as drought, flooding, chilling, and mechanical wounding. This "stress ethylene" leads to stress responses such as abscission, senescence and wound healing.

Physiological Effects

1. It breaks the seed and bud dormancy in some species. When applied to the seeds of cereals, ethylene breaks the dormancy and initiates germination. Ethylene treatment is sometimes used to promote sprouting in potato tubers and other bulbs.

2. Ethylene is capable of inducing adventitious root formation in leaves, stems, flower stems and even in other roots. But this response requires unusually high concentrations of ethylene (10 mL L^{-1}).

3. Senescence is a genetically programmed developmental process that affects all the tissues of the plant. Ethylene and cytokinin play an important role in leaf senescence. The exogenous application of ethylene or ACC (the precursor of ethylene) accelerates leaf senescence, while treatment with exogenous cytokinins delays leaf senescence.

Cellular and Molecular Modes of Action

Ethylene response involves binding to a receptor, followed by activation of one or more signal transduction pathways leading to the cellular response. Ultimately, ethylene exerts its effects by altering the pattern of gene expression. In recent years, much has been learned about the mechanism of ethylene action through the analysis of the ethylene response mutants in *Arabidopsis*.

One of the primary effects of ethylene is to alter the expression of various target genes. Ethylene increases the mRNA transcript levels of numerous genes, including the genes that encode cellulose, chitinase, β-1,3-glucanase, peroxidase, chalcone synthase (key enzyme in flavonoid biosynthesis; and a pathogenesis-related (PR) protein) as well as ripening-related genes and ethylene biosynthetic genes. Regulatory sequences called **ethylene response elements (EREs)** that confer ethylene responsiveness to a promoter have been identified from the ethylene-regulated PR genes.

Four proteins that bind to ERE sequences were identified in tobacco. These proteins are called **ERE-binding proteins (EREBPs)**. The genes that encode the EREBPs may represent ethylene primary response genes, the gene products of which may regulate the expression of secondary response genes. The steady-state level of these EREBPs increases dramatically following ethylene treatment.

Ripening-related gene expression has been studied extensively in developing tomato fruits, and several genes that are highly regulated during ripening have been identified. During tomato ripening, the fruit softens as a result of cell wall hydrolysis and changes from green to red as a consequence of chlorophyll loss and the synthesis of the carotenoid pigment lycopene. At the same time, aroma and flavour components are produced.

Analysis of mRNA from the wild type tomato fruits and the transgenic tomato plants that are genetically engineered to lack ethylene biosynthesis has revealed that gene expression during ripening is regulated by at least two independent pathways: an ethylene-dependent pathway and a developmental, ethylene-independent pathway. Genes involved in lycopene and aroma biosynthesis, respiratory metabolism, and ACC are transcriptionally regulated by ethylene, whereas genes encoding ACC oxidase and chlorophyllase are regulated by an ethylene-independent developmental programme. Interestingly, the transcription of the gene encoding polygalacturonase, an enzyme thought to be responsible for cell wall hydrolysis during ripening, is regulated by the ethylene-independent developmental programme, but its translation appears to be regulated by ethylene.

Commercial Uses

It is difficult to apply ethylene in the field as a gas, but this limitation can be overcome if an ethylene-releasing compound is used in increasing the agricultural produce. The most widely used such compound is ethephon or 2-chloroethyl phosphoric acid known by its trade name, **Ethrel**.

Ethephon is sprayed in aqueous solution and is readily absorbed and transported within the plant. It releases ethylene slowly by a chemical reaction, allowing the hormone to exert its effects.

The other system in which the effects of ethylene have been studied at the molecular genetic level is tomato fruit ripening. Ripening is controlled by ethylene, and a number of mutants showing delayed ripening have been isolated. The Never-ripe (Nr) mutant of tomato does not exhibit fruit ripening at all.

REVIEW QUESTIONS

1.   Write in detail the biosynthesis and metabolism of auxins.

2.   How are gibberellins synthesized by plants?

3.   What are the physiological roles of cytokinins?

4.   Describe the synthesis and mode of action of abscisic acid.

5.   What are the cellular and molecular modes of ethylene action?

REGULATION OF GENE EXPRESSION DURING PLANT DEVELOPMENT

Development is the sum of all changes that an organism goes through in its life cycle. The life cycle of any plant can be broadly classified into two phases of development—vegetative and reproductive phases. Vegetative phase starts from haploid gametes—egg and sperm—and ends with the diploid zygote, hence also referred to as gametophytic phase, whereas reproductive phase starts from the diploid zygote and ends with the meiotic phase of the spore mother cell to form gametes.

The duration of both phases differ with the organism and can be dependent or independent of one another. Development of a plant is dependent on the coordination of various factors and is subject to control at three distinct levels.

   i. Hormonal control

  ii. Environmental control

 iii. Genetic control

Hormonal control involves hormones (intrinsic factors) to control the development. The major hormones involved are auxin, gibberellin, cytokinins, ABA, and ethylene (Refer chapter 5). Environmental control involves extrinsic factors such as heat, temperature, water, etc. Genetic control involves the expression of particular genes at a particular time for proper development.

Development can also be referred to as an interaction of the genome with the cytoplasm and external environment to produce a programmed sequence of typically irreversible events.

PLANT DEVELOPMENT

Most of the genetic studies on development use the model plant *Arabidopsis*. It has been called the *Drosophila* of plant biology because of its widespread use in the study of plant genetics and molecular genetic mechanisms, particularly in an effort to understand plant developmental changes. It was the first higher plant to have its genome completely sequenced. *Arabidopsis thaliana* is a member of the family Brassicaceae, the mustard family (Figure 6.1). It is a small plant, well-suited for laboratory culture and experimentation.

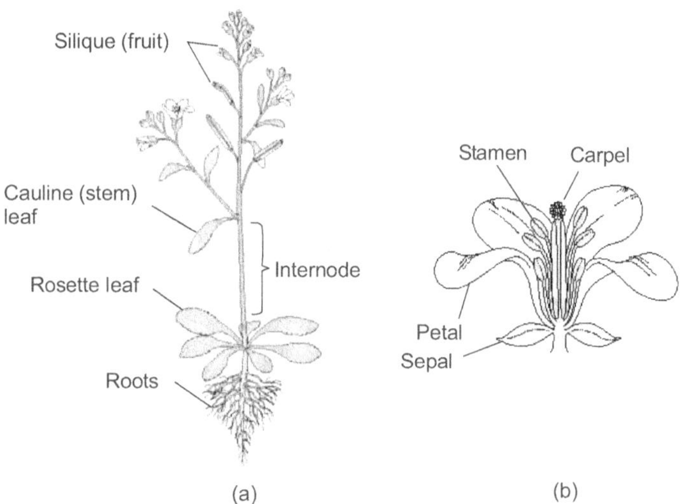

Figure 6.1 *Arabidopsis thaliana.* (a) Mature *Arabidopsis* plant showing the various organs. (b) Flower showing the floral organs.

Development of plants takes place in three stages:

1. Embryogenesis
2. Vegetative phase
3. Reproductive phase

EMBRYOGENESIS

Embryo is an immature plant formed after sexual or asexual reproduction. It is found in seeds and consists of an embryonic axis bearing a terminal bud (plumule), root (radicle) and one or more seed leaves (cotyledons). Embryogenesis is the development of a single-celled zygote into a multicellular embryo. A complete **embryo** has the basic body plan of the mature plant and many of the tissue types of the adult. This is one of the important stages of plant development. It does not generate the tissues and organs of the adult. Instead, the root and shoot apical meristems are specified, thus establishing the basic architecture of the seedling and the differentiation of vegetative tissue and organ systems.

The fact that a zygote gives rise to an organized embryo with a predictable structure tells us that the zygote is genetically programmed to develop in a particular way and that the various steps are tightly controlled during embryogenesis.

If these processes were to occur at random in the embryo, the result would be a clump of disorganized cells with no defined form or function. These studies have been carried out mostly in *Arabidopsis* assuming that most angiosperms probably use similar developmental mechanisms with subtle changes in time and place bringing about the diversity in form between the various flowering plants rather than by different mechanisms altogether (Doebley and Lukens, 1998).

In plants, as in all other eukaryotes, the union of one sperm with the egg forms a single-celled zygote. In angiosperms, however, this event is accompanied by a second fertilization event, in which another sperm unites with two polar nuclei to form the triploid endosperm nucleus, from which the **endosperm** (the tissue that supplies food for the growing embryo) will develop. This is termed as double fertilization which is unique to the flowering plants.

Two basic developmental patterns (Figures 6.2 and 6.3) during embryogenesis, which persist and can easily be seen in the adult plant are

1. The apical-basal, axial developmental pattern.
2. The radial pattern of tissues found in stems and roots.

VEGETATIVE DEVELOPMENT

The vegetative development consists of three stages.

1. Seed development and germination
2. Chlorophyll development of the vegetative plant (synthesis of chloroplast proteins and establishment of a whole plant)
3. Shoot and root development

Seed Development and Germination

Seeds grown for food generally contain large quantities of stored carbohydrate, protein and other reserves.

Seed	**Protein dry weight (%)**
Legumes	20–40
Cereals	7–16
Pea seeds	80

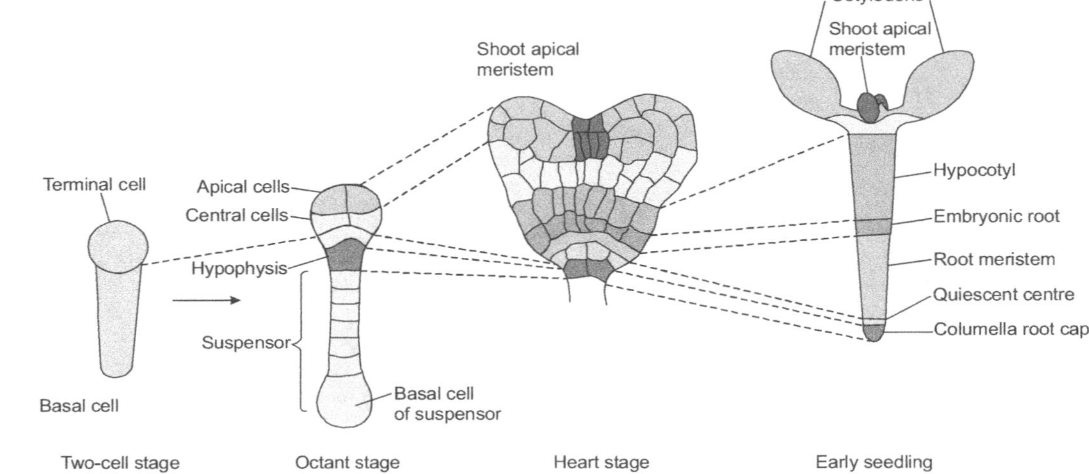

Figure 6.2 Apical-basal organization illustrating the origin of organs of the early *Arabidopsis* seeding from specific regions of the embryo.

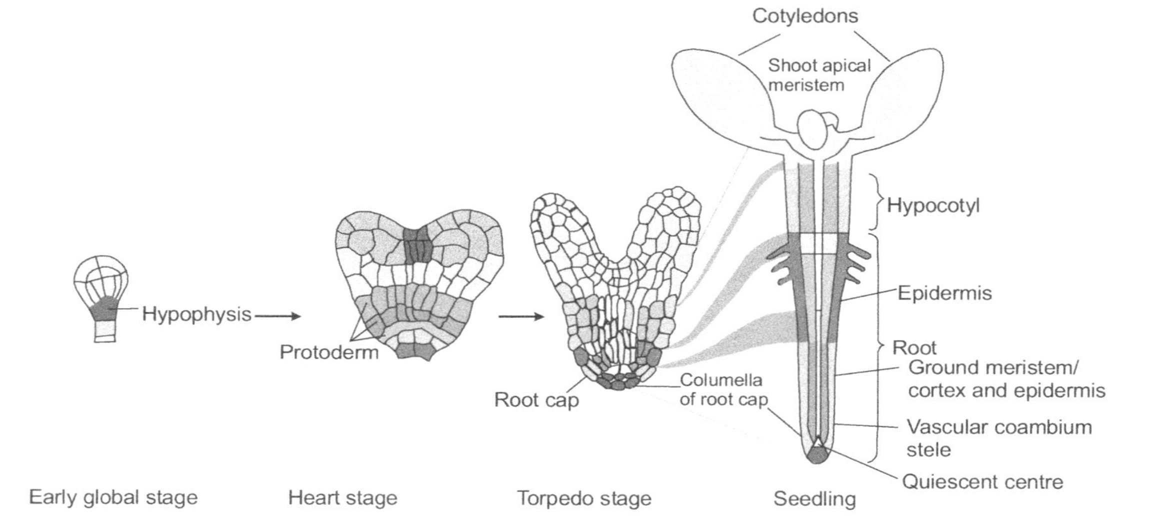

Figure 6.3 Radial tissue patterns illustrating the origin of different tissues and organs from embryo in *Arabidopsis*

Proteins are usually stored in the endoplasmic reticulum and vacuoles. In legumes, they are stored in the swollen cotyledons of the embryo. In cereals, the main reserve material is starch, which accumulates in the amyloplasts of the endosperm cells during grain filling. The growing embryo utilizes starch, proteins and other reserves stored in the endosperm during germination in order to develop into a young seedling. When barley grains imbibe water, metabolic activity quickly develops in the cells of the embryo and the aleurone layer. After a few days, mobilization of the stored reserves is quite noticeable and the endosperm becomes partly liquefied.

The following are the major proteins that are synthesized only during the development of the seeds.

Seed	**Storage proteins**
Pea	Vicilins and legumins
Barley	Hordeins
Wheat	Gliadins
Maize	Zeins

After setting a seed, it will either undergo a dormant period for a while or directly germinate into a seedling if the environment is favourable. During seedling development or germination, the seed needs energy which is obtained from the stored proteins (seed storage proteins). These proteins start accumulating between 10–20 days after flowering, accompanied by a phase of rapid cell division. During this period, the cotyledonary cells become polyploid and thus the number of copies of the storage protein encoding genes in the cells increase. In addition, half of the cellular mRNA is involved in storage protein synthesis.

In pea, vicilin begins to accumulate before legumin, and there are several types of genes for each protein. The relationship between mRNA content and protein synthesis breaks down during seed desiccation. Although protein synthesis stops, legumin mRNA remains at high levels.

Genetic regulation of reserved food material   Several conserved DNA sequences may be involved in regulating the expression of the genes. Examples are the 300-base-pair region upstream of the ATG codon of wheat α-gliadin and barley β-hordein. Similar sequences are present further upstream. A 15-bp conserved sequence is found upstream of the TATA box of all maize zein genes and these signals resemble the viral and animal "enhancer" core sequences, which stimulate gene expression. A protein of maize endosperm has been identified, that binds specifically to a 22-base pair region of maize zein gene between 339 and 318 nucleotides upstream of the start of transcription.

Genetic regulation during seed maturation and dormancy   After the complete development, the embryo enters into a dormant period. Dormancy is a living condition in which no growth occurs, especially under unfavourable conditions. This is brought about by the loss of water and a general shutting down of gene transcription and protein synthesis. It is regulated by specific gene expression (e.g.) ABI-3 (abscisic acid insensitive-3) and FUSCA 3. These genes are necessary for the initiation of dormancy and are sensitive to the hormone ABA, a signalling molecule that initiates seed and embryo dormancy. ABI-3 also controls the expression of genes encoding the storage proteins that are deposited in the cotyledons during the maturation phase of embryogenesis.

Hormonal regulation during germination   Transverse sectioning of the endosperm at different times after imbibition shows that the solubilization process begins at the outside, in association with the cells of the aleurone layer and progresses inwards. No solubilization occurs if the embryo is removed from the endosperm soon after the imbibition. These observations led to the demonstration of a diffusible stimulus moving from the embryo to the aleurone layer. This was later identified to be GA3. GA3 is one of the 80 naturally occurring gibberellins. It is thought that both GA1 and GA3 are active in the aleurone layer.

The gibberellins isolated from aleurone layer perform the following important functions.

i. They induce the production of new hydrolytic enzymes such as acid phosphatase, beta-1,3 glucanase, ribonuclease, alpha-amylase and protease.

ii. They bring about the establishment of mechanisms for the secretion and release of new and pre-existing enzymes to the outside of the cells.

iii. They help in the secretion of enzymes and synthesis of poly(A) containing mRNAs (e.g. alpha-amylase).

Genetic regulation during germination Alpha-amylase appears within one hour of GA3 application and continues to increase with time. Isoenzymes of alpha-amylase can be divided into two groups on the basis of isoelectric point and reaction with antibodies. In wheat, 3 families of alpha-amylase genes have been identified. *amy*3, present on chromosome 5, *amy*1 on chromosome 6 and *amy*2 on chromosome 7. *amy*3 group is expressed only in developing grains, whereas members of *amy*1 and 2 are expressed preferentially in aleurone layers in response to GA.

Development of Chloroplast

During seed germination and leaf growth in darkness, the proplastids increase greatly in size and develop into etioplasts which are several μm long.

Proplastids　These contain large amounts of internal membranes including a characteristic prolamellar body formed by a *p*-crystalline lattice of tubules.

Etioplastids　These contain enzymes involved in CO_2 fixation and metabolism and some of the proteins normally found in the thylakoids of chloroplast. They lack chlorophyll and some other components required for photosynthesis.

Transformational changes following illumination　The responses of etioplasts to light are the following.

- ✖ Development of the chlorophyll precursor and the protochlorophyllide *a*

- ✖ Formation of a blue light receptor

- ✖ The photoconvertible pigment phytochrome that remains in P_r form after a flash with red light (660 nm) and occurs as the active P_{fr} when it senses a pulse of far-red light (730 nm)

Ultrastructural changes following illumination　The following ultrastructural changes convert etioplastids to chloroplast.

- ✖ Dispersal of the prolamellar body of etioplasts.

- ✖ An increase in the amount of thylakoid membrane.

- ✖ Rapid formation of small amount of chlorophyll *a* by the light-stimulated reduction of protochlorophyllide *a* to chlorophyllide *a* by the thylakoid enzyme NADPH-protochlorophyllide oxidoreductase, which in most plants requires light in order to reduce its substrate. The chlorophyllide *a* is then esterified with geranyl geranoil before reduction to produce the phytol side chain of chlorophyll *a*.

✗   Photoactivation of phytochrome results in the subsequent stimulation of the chlorophyll biosynthetic pathway.

✗   Accumulation of chlorophyll-binding proteins.

✗   Synthesis of more membrane material.

✗   Formation of grana from thylakoids.

✗   Start of biochemical activity (one thylakoid protein declines in quantity and many more appear).

✗   Synthesis of RuBisCO, other Calvin cycle enzymes and thylakoid proteins (although some of them are present in small quantities in dark).

Regulation of gene expression during chlorophyll development   Light has effects on genes of both nucleus and plastids. An example is the *psb*A gene that encodes a 32-kDa product, the atrazine-binding QB protein. This was originally known as the photo gene. When etiolated plants of maize, spinach, mustard and pea are transferred to light, there is a large increase in the amount of *psb*A transcripts which is due to the activation of the phytochrome system.

*rbc*L, *atp*A, B, E, H, *psa*A1, A2, *psb*B, D are other examples. Many of these mRNAs are also synthesized in the dark and the extent of light stimulation depends on the developmental stage of the plant when illuminated.

It may be influenced by a number of factors such as,

1.   Number of plastids/cells.

2.   Number of copies of the plastome/plastid.

3.   The rate of division and expansion of the cells in the leaf.

The rate of transcription of chloroplast-coded genes increase 2–3-fold after illumination, which is insufficient to explain the observed differences in the amounts of proteins accumulating exclusively in light and dark.

Thus, it is believed that the rates of mRNA degradation, translational control of mRNA and differences in turnover rate of special proteins play an important role in the development of chlorophyll in the presence of light. Plants grown in dark frequently synthesize RuBisCO, which gets accumulated in etioplasts. Its synthesis is enhanced by light.

The transcription rate of *rbcS* genes located in the nucleus is 18 times higher following illumination, whereas rRNA genes are transcribed only twice effectively. *rbcS*-3A and *rbcS*-3C are highly expressed in green leaves and account for 40% and 34% of the total *rbcS* mRNA respectively.

A flash of red light stimulates the accumulation of transcripts from 3A and 3C in etiolated leaves, but has much less effect on the *rbcS-E9* and *rbcS-80* genes. In mature leaves, the blue light receptor also seems to be important in regulating *rbcS* gene transcription.

Very little mRNA of Cab proteins of light-harvesting complex II is observed in the leaves of many etiolated plants. The mRNA accumulates rapidly in response to photoactivation of the phytochrome system by red light. *cab* genes are much more sensitive to light than *rbc* genes, requiring a 1000-fold lower red light fluence rate for stimulating mRNA accumulation.

The transcription of *cab* genes in wheat follows a diurnal rhythm, with peak activity in the light and very little mRNA synthesis in the dark. Although the mRNAs appear after illumination, the light-harvesting proteins do not accumulate unless chlorophyll synthesis occurs. In the absence of chlorophyll, the proteins are unstable and are degraded.

If chlorophyll is present, the proteins are inserted into the thylakoids where they combine with chlorophyll *a* and *b*.

cis-acting *DNA* sequences with light and organ-specific expression of *rbcS* and *cab* genes have been identified. A few examples are given below.

1. 33-bp sequence upstream of TATA box was found to be important for light-regulated expression.

2. Transcription was enhanced by DNA sequences between −1052 and −352 from the transcription start site.

3. A 410-bp 5′-upstream sequence from *rbcS-3A* was sufficient to obtain red- and blue-light-induced expression in the leaves of regenerated plants.

4. A 352-bp upstream region sufficed for the *rbcS-E9* gene.

5. Sequences from −327 to −48 of *rbcS-E9* and −410 to +15 of *rbcS-3A* have two important features. They resemble enhancers in that they can function in either orientation. They contain silencer sequences that repress gene expression in roots. These genes have 4 conserved regions in addition to the TATA box which are thought to contain light-responsive elements (LREs) that are required for the expression of *rbcS* genes.

Green *et al.* identified a protein factor from pea nuclei called GT-1 that binds to these putative regulatory sequences upstream of the *rbcS*-3A gene. The two GT-1 binding sequences, which are also repeated further upstream, contain G residues identified as important for interaction with the protein. The GT-1 protein is present in the leaves of pea grown in both dark and light. Therefore, in order to regulate expression, it must either be post-translationally modified or sterically prevented from binding by another factor.

In *rbcS* gene family (Petunia), it has been shown that two of the 8 genes show quite different levels of expression, with one giving 100-fold greater mRNA levels than the other. Transgenic Petunia containing chimeric genes indicates that sequences downstream from the protein synthesis termination codon are responsible for this differential expression. In *cab*-1 gene of wheat, a 268-bp enhancer sequence located between –357 and –89 from the start of transcription, conferred phytochrome regulation. The 5′ deletion analysis showed that the CAAT sequence could be dispensed with for phytochrome-regulated expression.

Interestingly, there is no obvious homology between the LREs from *rbcS* and *cab* genes. This suggests that different *trans*-acting factors may regulate the two genes. This may be related to the different red light fluence rates required for stimulating mRNA accumulation.

Shoot and Root Development

The development of a whole plant from a seed involves the active division of meristems. Meristems are populations of small, isodiametric cells with embryonic characteristics. Vegetative meristems are self-perpetuating and these undifferentiated cells that retain the capacity for cell division indefinitely are called the stem cells.

Meristems can be classified into primary and secondary based on their origin. The root and shoot apical meristem formed during embryogenesis are called primary meristems and meristems developed during post-embryonic development are said to be secondary meristems. Based on its position and function, secondary meristem can be classified as

1. Axillary meristem—formed in the axils of plants.

2. Intercalary meristem—found within organs.

3. Branch root meristem.

4. Vascular cambium present in the vascular system and responsible for the development of wood.

5. Cork cambium developing within mature cells of the cortex and the secondary phloem.

Vegetative meristem may be converted directly into floral meristems (inflorescence meristems) when the plant is induced to flower. Floral meristems differ from vegetative meristems in that instead of leaves they produce floral organs: sepals, petals, stamens and carpels. Floral meristems are determinate and inflorescence meristems are indeterminate.

Shoot apical meristem generates the stem as well as the lateral organs attached to the stem, e.g. leaves and lateral buds. The shoot apical meristem is located at the extreme tip of the shoot but it is surrounded and covered by immature leaves, which are produced by the activity of the meristem. The shoot apex consists of the apical meristem plus the most recently formed leaf primordia. The leaves of most plants are the organs of photosynthesis. This is where light energy is captured and used to drive the chemical reactions that are vital to the life of the plant.

Roots grow and develop from their distal ends. Roots are adapted for growing through soil and absorbing the water and mineral nutrients in the capillary spaces between soil particles. Four developmental zones have been distinguished in a root tip:

✖ the root cap,

✖ the meristematic zone,

✖ the elongation zone and

✖ the maturation zone.

The root cap protects the apical meristem from mechanical injury as the root pushes its way through the soil. The meristematic zone lies just under the root caps and it generates the primary root. The elongation zone is the site of rapid and extensive cell elongation, and the maturation zone is the region in which cells acquire their differentiated characteristics. Cells enter this zone after division and elongation have ceased. The radial pattern of differentiated tissues becomes obvious in the maturation zone.

The root apical meristem of *Arabidopsis* has the following structure.

The quiescent centre is composed of a group of four cells that usually do not divide after embryogenesis. The cortical-endodermal stem cells form a ring of cells that surround the quiescent centre. These stem cells generate the cortical and endothermal layers. The columella stem cells are the cells immediately above (apical to) the central cells. They divide anticlinally (transverse) and periclinally (longitudinal) to generate a sector of the root cap known as the columella. The root cap-epidermal stem cells are similar to the tier of the columellar stem cells but form a ring surrounding them and these cells produce the lateral root cap.

The stelar cells are a tier of cells just behind the quiescent centre cells. These cells generate the pericycle and vascular tissues.

Protomeristem is the structure that becomes the root or shoot meristem upon germination. The shoot apical meristem of *Arabidopsis* has the following structure.

The shoot apical meristem (SAM) is a population of cells located at the tip of the shoot axis. It produces lateral organs and stem tissues, and regenerates itself. The primordia can be divided into regions called the central zone (CZ), peripheral zone (PZ) and rib zone (RZ). The central zone contains a core of stem cells, while the peripheral zone is the site of production of lateral organ primordia and the rib zone gives rise to differentiated cells of the growing stem (Figure 6.4). Further development forms distinct layers at the primordium (L1, L2 and L3).

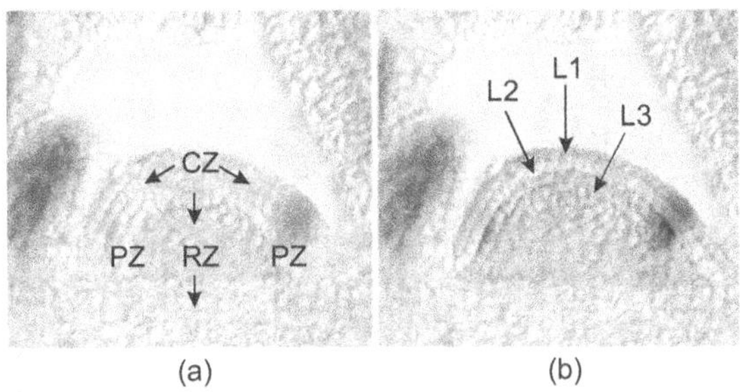

Figure 6.4   Architecture of the *Arabidopsis* shoot meristem

Table 6.1 lists some of the genes involved in the different stages of vegetative development of the plant.

REPRODUCTIVE DEVELOPMENT

It consists of three stages. They are

- ✗ Flowering
- ✗ Pollination
- ✗ Fertilization

Flowering

Always there is a strong correlation between flowering and seasons. The flowering phenomenon poses fundamental questions like:

1. How do plants keep track of the seasons of the year and the time of day?

2. Which environmental signals control flowering and how are those signals perceived?

3. How are environmental signals transduced to bring about the developmental changes associated with flowering?

Table 6.1 Genes involved in the regulation of gene expression

Gene	Function	Mutated gene
GNOM gene	Axial patterning.	Embryos become spherical and lack axial polarity entirely.
Monopteros gene (*mp* gene)	Required for the formation of the primary root and vascular tissue in postembryonic development.	Lack both a hypocotyl and a root. Apical structures are not structurally normal. Tissues of the cotyledons are disorganized. Show abnormalities at the octorut storage.
Short root and scarecrow genes (*shr* and *scr*)	Function in the establishment of the radial tissue pattern in the root and hypocotyl during embryogenesis. Required for the maintanence of radial pattern during postembryonic development.	Show defects in the radial tissue patterning. Produce roots with a single-celled layer of ground tissue, which show characteristics of both endodermal and cortical cells. Lack a layer of starch sheath, a structure that is involved in the growth response to gravity.

| HOBBIT gene (*hbt* gene) | Important for the development of a functional root apical meristem. | Hypophysis does not form and the root meristem that subsequently forms lacks a quiescent centre and the columella. Appears to have root meristem, but it does not function when the seedlings germinate. Embryos are unable to form lateral roots. |
| Shoot meristemless gene (*stm* gene) | Required for the formation of the shoot promeristem and appears to suppress cell differentiation ensuring that the meristem cells remain undifferentiated. Therefore, the *stm* gene is necessary not only for the formation of the embryonic shoot apical meristem, but also for the maintenance of shoot apical meristem identity in the adult plant. | Mutants do not form a shoot apical meristem. |

The transition of shoot apex into flower or inflorescence involves major changes in the pattern of morphogenesis and cell differentiation at the shoot apical meristem. Ultimately this process leads to the production of the floral organs: sepals, petals, stamens and carpels. The events occurring in the shoot apex that specifically commit the apical meristem to produce flowers are collectively referred to as floral *evocation*. The developmental signals that bring about floral evocation include endogenous factors such as circadian rhythms, phase change, and hormones and external factors such as day length (photoperiod) and temperature (vernalization). In the case of photoperiodism, transmissible signals from the leaves, collectively referred to as the floral stimulus, are translocated to the shoot apical meristem. The interactions of these endogenous and external factors enable plants to synchronize their reproductive development with the environment. As plants initiate reproductive development, the vegetative meristem is transformed into an indeterminate primary inflorescence meristem that produces floral meristems on its flanks.

Floral meristems initiate four different types of floral organs: sepals, petals, stamens and carpels. These sets of organs are initiated in concentric rings called whorls (Figure 6.5) around the flanks of the meristem.

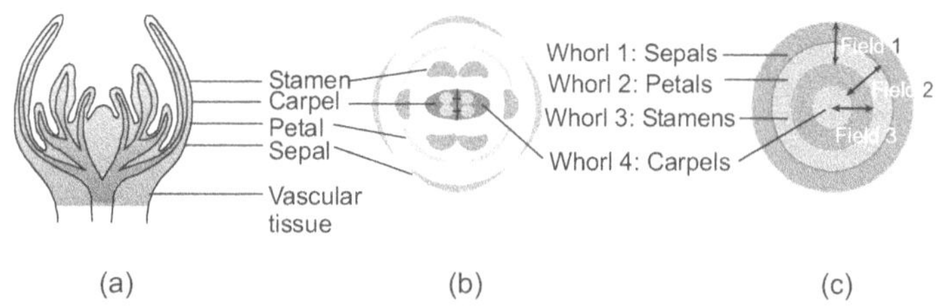

Figure 6.5 (a) Longitudinal section through developing flower (b) Cross-section of developing flower showing floral whorls (c) Schematic diagram of developmental fields.

Genes that regulate floral development Mutations have identified three classes of genes that regulate floral development: floral organ identity genes, cadastral genes and meristem identity genes.

(a) *Floral organ identity genes* These genes directly control floral identity. The proteins encoded by these genes are transcription factors that likely control the expression of other genes whose products are involved in the formation and/or function of floral organs.

The genes that determine floral organ identity were discovered as floral homeotic mutants, for example, *Arabidopsis* plants with mutations in the APETALA 2 (*ap2*) gene produce flowers with carpels where sepals should be, and with stamens where petals normally appear. The homeotic genes that have been cloned so far encode transcription factors—proteins that control the expression of other genes. Most plant homeotic genes belong to a class of related sequences known as MADS box genes, whereas animal homeotic genes contain sequences called homeoboxes. MADS box genes share a characteristic, conserved nucleotide sequence known as a MADS box, which encodes a protein structure known as the MADS domain. The MADS domain enables these transcription factors to bind to DNA that has a specific nucleotide sequence.

Five different genes are known to specify floral organ identity in *Arabidopsis*.

APETALA 1 (*ap1*), APETALA 2 (*ap2*), APETALA 3 (*ap3*), PISTILLATA (PI) and AGAMOUS (AG). The organ identity genes initially were identified through mutations that dramatically alter the structure and thus the identity of the floral organs produced in two adjacent whorls. For example, plants with the *ap2* mutation lack sepals and petals. Plant bearing *ap3* or *pi* mutations produce sepals instead of petals in the second whorl and carpels instead of stamens in the third whorl. Plants homozygous for the *ag* mutation lack both stamen and carpels.

Since mutation in these genes change the floral organ identity without affecting the initiation of flowers, they are said to be homeotic genes. These homeotic genes fall into three classes, type A, B and C, defining three different kinds of activities (Figure 6.6).

1. Type A activity, encoded by *ap1* and *ap2*, controls organ identity in the first and second whorls. Loss of type A activity results in the formation of carpels instead of sepals in the first whorl, and of stamens instead of petals in the second whorl.

2. Type B activity, encoded by *ap3* and *pi*, controls organ determination in the second and third whorls. Loss of type B activity results in the formation of sepals instead of petals in the second whorl, and of carpels instead of stamens in the third whorl.

3. Type C activity, encoded by *ag*, controls events in the third and fourth whorls. Loss of type C activity results in the formation of petals instead of stamens in the third whorl and replacement of the fourth whorl by sepals.

The role of the organ identity genes in floral development have been dramatically illustrated by experiments in which two or three activities are eliminated (*ap1*, *ap2*, *ap3/pi* and *ag*) to produce floral meristems that develop as pseudoflowers; all the floral organs are replaced with green leaf-like, whorled phyllotaxy.

(b) *Cadastral genes*   These genes act as spatial regulators of the floral organ identity genes by setting boundaries for their expression.

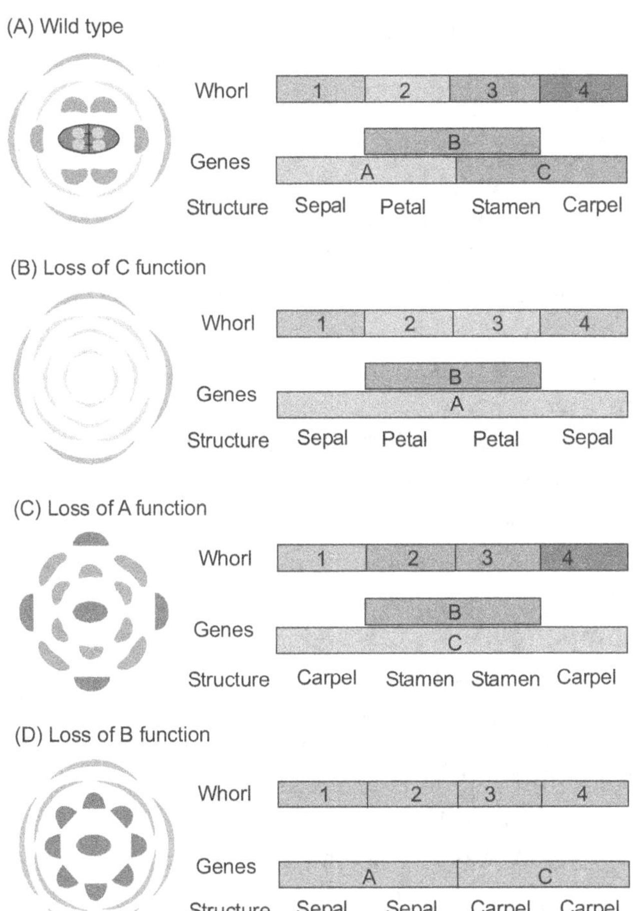

Figure 6.6 Interpretation of the phenotypes of floral homeotic mutants based on the ABC model. (a) Wild type (b) Loss of C function results in expansion of the A function throughout the floral meristem. (c) Loss of A function results in the spread of C function throughout the meristem. (d) Loss of B function results in the expression of only A and C functions.

(c) *Meristem identity genes* These genes are necessary for the initial induction of the organ identity genes. These genes are the positive regulators of floral organ identity.

Meristem identity genes must be active for the primordia formed at the flanks of the apical meristem to become floral meristems. In *Arabidopsis*, AGAMOUS-LIKE 20 (*agl20*), APETALA 1 (*ap1*) and leafy (*lfy*) are all critical genes in the genetic pathway that must be activated to establish floral meristem identity. *agl20* plays a critical role in floral evocation by integrating signals from several different pathways involving both environmental and internal cues. *agl20* thus appears to serve as a master switch initiating floral development. Once activated, *agl20* triggers the expression of *lfy*, and *lfy* turns on the expression of *ap1*.

Phytoperiodism and vernalization are two of the most important mechanisms underlying seasonal responses. Photoperiodism is a response to the length of the day, vernalization is the promotion of flowering at subsequent higher temperatures, brought about by exposure to cold. Other external cues are total light radiation, and water availability. The evolution of both internal (autonomous) and external (environmental sensing) control systems enables plants to carefully regulate flowering at the optimal time for reproductive success.

Biochemical signalling involved in flowering Some of the events that result in floral evocation are triggered by biochemical signals arriving at the apex from other parts of the plant, especially from the leaves. This biochemical signal is said to be a universal flowering hormone, named as florigen by Mikhail Chailakhyan in 1930s. Gibberellins and ethylene can also induce flowering in *Arabidopsis* by activating expression of the *lfy* gene.

Gene regulation involved in flowering Recent genetic studies have established that there are four genetically distinct developmental pathways that control flowering in *Arabidopsis*. There are four developmental pathways (Figure 6.7).

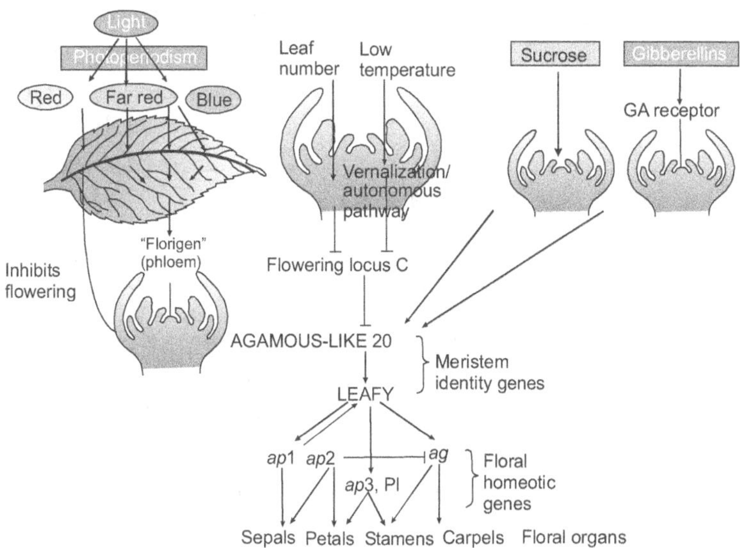

Figure 6.7 Four developmental pathways for flowering in *Arabidopsis*: the photoperiodism, autonomous/ vernalization, sucrose, and gibberellin pathways. A transmissible floral stimulus ("florigen" from leaves) is involved only in the photoperiodic pathway (After Blazquez, 2000).

They are the photoperiodic, autonomous/vernalization, sucrose and GA (Gibberellic acid) pathways. All of these pathways converge to regulate the meristem identity genes AGAMOUS-LIKE 20 (*agl20*) and Leafy (*lfy*), which in turn regulate the floral homeotic genes to produce the floral organs. The existence of multiple pathways for flowering provides angiosperms with the flexibility to reproduce under a variety of environmental conditions, thus increasing their evolutionary fitness.

Fruit Development and Ripening

Earlier it was thought that ripening was a degradative process brought about by a breakdown in "organizational resistance" of the cell. The fact that mutations were interfering with the ripening process made it clear that specific genes were involved in ripening. Indeed it has been now demonstrated that ripening and senescing organs synthesize mRNAs encoding enzymes that bring about developmental changes.

Fruits are divided into climacteric and non-climacteric depending on their pattern of ripening.

Climacteric fruits are characterized by an increased respiratory rate and ethylene synthesis during ripening whereas non-climacteric fruits show no respiratory rise during ripening. Examples of climacteric fruits are apples, pears, bananas and tomatoes. Non-climacteric fruits include oranges, lemons, grapes and strawberries.

Experiments with tomatoes show that an increase in natural ethylene synthesis precedes the respiratory climacteric. The application of ethylene increases its respiration. Aminoethoxyvinyl glycine is an inhibitor of ethylene synthesis. Silver and norbornadiene delay perception or inhibit ripening.

The functions of ethylene are as follows:

1. It stimulates senescence of leaves, flowers and petals.
2. It stimulates shedding of leaves and flowers at abscission zones.
3. It stimulates ripening of certain fruits.

Role of ethylene in stimulating ripening   Ethylene is synthesized with the help of ACC synthase and ACC oxidase. Ethylene stimulates its own synthesis and also promotes ripening by causing the accumulation of specific mRNAs.

Methionine → S-adenosylmethionine → 1-aminocyclo-propane1-carboxylic acid → ethylene receptor → mRNA and enzyme synthesis → ripening changes.

ACC-1-aminocyclopropane 1-carboxylic acid.

Once initiated, ethylene synthesis is autocatalytic. Ethylene is used commercially to ripen bananas that are picked and transported unripe. An inhibitor of ethylene perception, silver is used to prolong the vase life of carnations and other flowers.

Changes occurring during ripening

- ✗ Rise in respiration
- ✗ Softening of the cell walls
- ✗ Synthesis of compounds that contribute to flavour and aroma
- ✗ Conversion of chloroplasts to chromoplasts
- ✗ The production of chromoplasts involves the degradation of the chlorophyll and starch and dismantling of the thylakoids
- ✗ Decline in the expression of genes encoding photosynthetic enzymes

Genetic regulation during ripening Nineteen mRNAs appear to increase greatly in quantity during ripening. For example, the wall-softening enzyme, polygalacturonase exists in three-isoenzyme forms which are structurally and immunologically related. Polygalacturonase (PG) is absent in green fruits and is synthesized de novo during ripening.

Nr mutant: "Neverripe" mutant, which softens slowly, makes much less PG than normal.

Rin mutant: This has a mutation on chromosome 5 and makes only a trace of enzyme. It shows very little softening.

PG gene is located on chromosome 10. It plays a role in degrading the pectin fraction of the cell wall during softening. PG isoenzymes have been shown to degrade the cell walls of mature green fruit *in vitro*.

During normal ripening, the synthesis of PG is preceded by ethylene production and treatment of mature unripe fruit with ethylene induces ripening changes including PG synthesis.

PG is synthesized initially with a 71-amino acid presequence that may be involved in transport and secretion across the plasma membrane into the cell wall.

Ethylene stimulates major increases in a number of ripening-related mRNAs. This occurs before ripening changes become visible and is consistent with the idea that the mRNAs cause ripening.

It has been estimated that the increase in PG mRNA concentration during ripening is of the order of 1000-fold. The amount of PG mRNA found in the rin mutant is 100-fold less than the normal level due to the lack of softening. Experiments indicate that the rin mutant does contain the PG gene but does not express it at higher levels. This is consistent with the suggestion that rin is a mutation in the regulatory apparatus that normally switches on the PG gene. The mapping of PG to chromosome 10 and rin to chromosome 5 is consistent with the lesion involving a *trans*-acting factor, but it is not known whether the mutation affects a DNA-binding protein. Although PG mRNA does accumulate in mature green fruit treated with ethylene, the response is not as rapid as it is for other mRNAs.

It has been suggested that PG mRNA does not respond directly to ethylene, but some additional steps are required to elicit the response. Another possibility is that the production of some mRNAs occurs at very low ethylene concentrations whereas higher levels of the gas are required to stimulate the appearance of others.

The perception of ethylene is required for continued PG mRNA production, since application of silver after ripening caused the PG mRNA to disappear.

Recently, a 1450-bp DNA fragment 5′ to the PG coding region has been shown to direct the synthesis of a CAT reporter gene in the fruit of transgenic tomato plants. This indicates that *cis*-acting control signals for ripening specific expression are located within this DNA region.

Ethylene induces the expression of a number of genes that are involved in fruit ripening and senescence with the help of ACC. Treatment of plants with chemical compounds that block ethylene production delays both fruit ripening and senescence. Thus premature fruit ripening might be prevented by inhibiting the plants' ability to synthesize ethylene. This can be achieved in several different ways. For example, transgenic plants that have been engineered to contain antisense RNA versions of either ACC synthase or ACC oxidase, whose activities are essential for the synthesis of ethylene, have much lower levels of ethylene and therefore the fruit that is produced by those plants has an extended storage life.

DIFFERENTIATION

Differentiation is the process by which a cell acquires metabolic, structural and functional properties that are distinct from those of its progenitor cell. In plants, unlike animals, cell differentiation is frequently reversible, (except sieve tube elements of phloem and tracheary elements of xylem), particularly when differentiated cells are removed from the plant and placed in tissue culture. This ability to dedifferentiate demonstrates that differential plant cells retain all the genetic information required for the development of a complete plant, properly termed **totipotency**.

REGULATION OF DEVELOPMENT BY CELL-TO-CELL SIGNALLING

Neighbouring cells and distant tissues and organs provide positional information. Cells in multicellular plants usually are in close contact with others around them, and the behaviour of each cell is carefully coordinated with that of its neighbours throughout the life of the plant. Furthermore, each cell occupies a specific position within the tissue and organ to which it belongs. Coordination of cellular activity requires cell–cell communication. A given gene or set of genes can exert an effect on development in neighbouring cells or even cells in distant tissues through cell–cell communication, via at least three different mechanisms.

1. Ligand-induced signalling
2. Hormonal signalling
3. Signalling via trafficking of regulatory proteins and/or mRNAs

Ligand-induced Signalling

There are evidences that cell wall components, particularly a class of glycoprotein macromolecules, known as **arabinogalactan proteins**, or **AGPs**, may communicate positional information that will determine a cell's fate. AGPs would not be involved in signalling over a distance, but rather in telling a given cell who its neighbours are. That information then would programme the cell to differentiate, or acquire a fate appropriate to its position.

Because plants have numerous, perhaps hundreds of, receptor kinases, many signalling events are expected to be initiated by ligand-induced protein phosphorylation.

The *CLV3* gene for example is expressed in cells of the L1 and L2 layers in the central zone of the shoot apical meristem, but not within the L3 layer or in the peripheral zone (Figure 6.8).

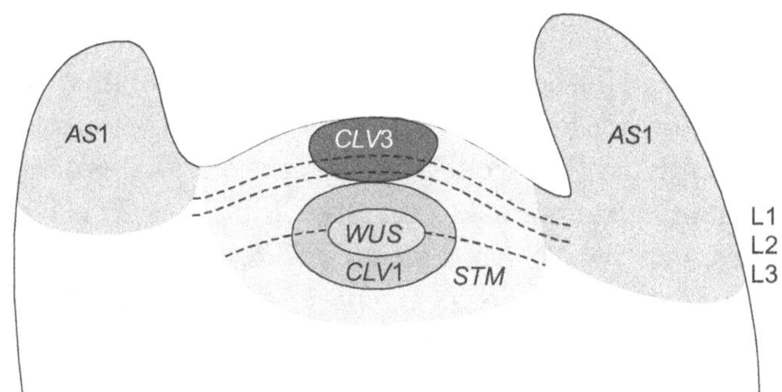

Figure 6.8   Patterns of expression of some developmentally important genes in the *Arabidopsis* shoot apical meristem.

In contrast, *CLV1* is expressed in deeper layers within the central zone in the L3 layer, as is the *WUS* gene. However, *CLV1* is expressed within a somewhat larger domain than *WUS*. Although *WUS* gene expression is required to maintain stem cell identity, *WUS* is expressed in only a small number of cells in the L3 layer of the meristem. It functions non-autonomously, acting on cells a short distance from the cells that express the gene. The CLV3 protein controls the size of the stem cell population in the shoot apex by negatively regulating the expression of *WUS* in the L3 layer. When *CLV1* or *CLV3* is knocked out by mutation, *WUS* gene expression spreads, and the number of undifferentiated stem cells expands (Brand *et al.*, 2000).

Hormonal Signalling

The plant hormones such as auxin, ethylene, gibberellins, abscisic acid, cytokinins, and brassinosteroids play roles in regulating development. These roles will be presented in some detail in the chapters devoted to these topics. However, auxin signalling is explained here as an example of the types of mechanisms these roles might entail.

Auxin signalling is essential for the development of axial polarity and the development of vascular tissue. Auxin has long been known to be the signal for the initiation of vascular tissue differentiation. This conclusion is based on studies of the effects of applied auxins and auxin transport inhibitors. Two *Arabidopsis* genes, namely *GNOM* and *MONOPTEROS* known to be essential for the development of axial polarity and tissue differentiation during embryogenesis and adult plant development, have been found to be involved in auxin signalling.

GNOM encodes a guanine nucleotide exchange factor that is a component of the cellular machinery that establishes cell polarity. This machinery, and the GNOM protein in particular, are required for the correct localization of the auxin efflux carrier protein PIN1 at the basal end of the procambium cells during the globular stage of embryogenesis and subsequently in vascular cells throughout development (Steinmann *et al.*, 1999; Grebe *et al.*, 2000).

Mutations in the *MONOPTEROS* (mp) gene result in seedlings that lack both hypocotyls and roots, although they do produce an apical region.

Thus, auxin is likely to be required for signalling early events necessary for organogenesis from the shoot apical meristem.

Signalling Via Trafficking of Regulatory Proteins and/or mRNAs

Symplastic communication between plant cells occurs via the plasmodesmatal connections through their cell walls. Most living cells in a plant are connected symplastically to their neighbours by plasmodesmata that pass through the adjoining cell walls and provide some degree of cytosolic continuity between them. There is increasing evidence that the signals exchanged through plasmodesmata include both regulatory proteins and mRNAs

(Zambryski and Crawford, 2000). The importance of plasmodesmata for cell–cell communication during development became apparent with the discovery that the mRNA of the maize meristem identity gene *KN1* cannot be detected in the L1 layer of the maize vegetative shoot apical meristem. The *KN1* gene is expressed only in cells of the L2 layer. The KN1 protein, however, is detected in all regions of the shoot apical meristem, including the L1 layer. Since the KN1 protein is not synthesized in the L1 layer, it must be transported into the L1 layer from the L2 layer, through the plasmodesmata joining them (Lucas *et al.*, 1995).

REVIEW QUESTIONS

1. What are the various factors controlling transcription of genes?

2. Discuss the regulation of expression of genes during the development.

PHYTOCHROME

INTRODUCTION

Light controls the normal development of a plant. There is a change in the colour of the plants grown in dark and light. Dark-grown plants are usually pale and tall with thin stems whereas light-grown plants are dark green and short with thick stems and more branches. When these dark-grown plants are transferred to places with direct sunlight, changes in their colour and growth were observed. This shows that light plays an important role in plant growth.

The growth in dark-grown plants is known as **etiolated growth**. In the absence of light, the seedling uses primarily stored reserves for etiolated growth. Light brings some initial rapid changes in etiolated plants and is called **photomorphogenesis**. Among the different pigments that can promote photomorphogenic responses in plants, the most important are those that absorb red and blue light, named **phytochrome.**

Definition

Phytochrome is a plant growth regulating photoreceptor protein that absorbs primarily red light and far-red light, and also blue light. It is a holoprotein that contains the chromophore phytochromobilin (a linear tetrapyrrole chromophore of phytochrome).

BIOCHEMISTRY OF PHYTOCHROME

Phytochrome is a blue protein pigment with a mass of 250 kDa. The existence of phytochrome in two forms came through the study on the germination of lettuce seeds by red and far-red light. Lettuce seeds were exposed to alternating treatments of red and far-red light. Nearly 100% of the seeds that received red light as the final treatment germinated; seeds that received far-red light as the final treatment failed to germinate (Flint 1936).

The result of these experiments gave two possible interpretations. One is that there are two pigments, a red-light-absorbing pigment and a far-red-light-absorbing pigment. Another possibility is the two pigments of seed germination. Alternatively there might be a single pigment that can exist in two interconvertible forms; a red light-absorbing form and a far-red light-absorbing form (Borthwick *et al.*, 1952).

The properties of phytochrome are

1. Phytochrome can interconvert between P_r and P_{fr} forms.
2. P_{fr} is the physiologically active form of phytochrome.
3. Phytochrome is a dimer composed of two polypeptides.
4. Phytochromobilin is synthesized in plastids.
5. Both chromophore and protein undergo conformational changes.
6. Two types of phytochromes are present.
7. Phytochrome is encoded by a multigene family.

In dark-grown or etiolated plants, phytochrome is present in a red-light-absorbing form, named as P_r, appears blue coloured, is converted by red light to a far-red-light-absorbing form called P_{fr}, appears blue-green. P_{fr} can be converted back to P_r by far-red light (Figure 7.1). This property is known as photoreversibility, and can be represented as

$$P_r \underset{\text{Far-red light}}{\overset{\text{Red light}}{\rightleftharpoons}} P_{fr}$$

In addition to absorbing red light, both forms of phytochrome absorb light in the blue region of the spectrum. The photoconversions of P_r to P_{fr} to P_r, are not one-step processes, but involve intermediates which are short-lived. These intermediates play a role in initiating or amplifying phytochrome responses under natural sunlight.

Phytochrome is a soluble protein with a molecular mass of about 250 kDa. It is a dimer made up of two equivalent subunits (Figure 7.2). Each subunit consists of two components, a light-absorbing pigment molecule called the chromophore, and a polypeptide chain called the apoprotein (125 kDa). Together, the apoprotein and its chromophore make up the holoprotein. The chromophore of phytochrome is a linear tetrapyrrole termed phytochromobilin. There is only one chromophore per monomer of apoprotein and it is attached to the protein through a thioether linkage to a cysteine residue (Figure 7.1).

Light can be absorbed only when the polypeptide is covalently linked with phytochromobilin to form the holoprotein. Phytochromobilin is synthesized inside plastids and is derived from 5-amino levulinic acid via a pathway that branches from the chlorophyll biosynthesis. It is transported to the cytosol by a passive process. Upon absorption of light, the P_r chromophore undergoes a *cis-trans* isomerization of the double bond between carbon 15 and 16 and rotation of the C_{14}-C_{15} single bond. During this conversion of P_r to P_{fr}, the protein moiety of the phytochrome holoprotein also undergoes a conformational change.

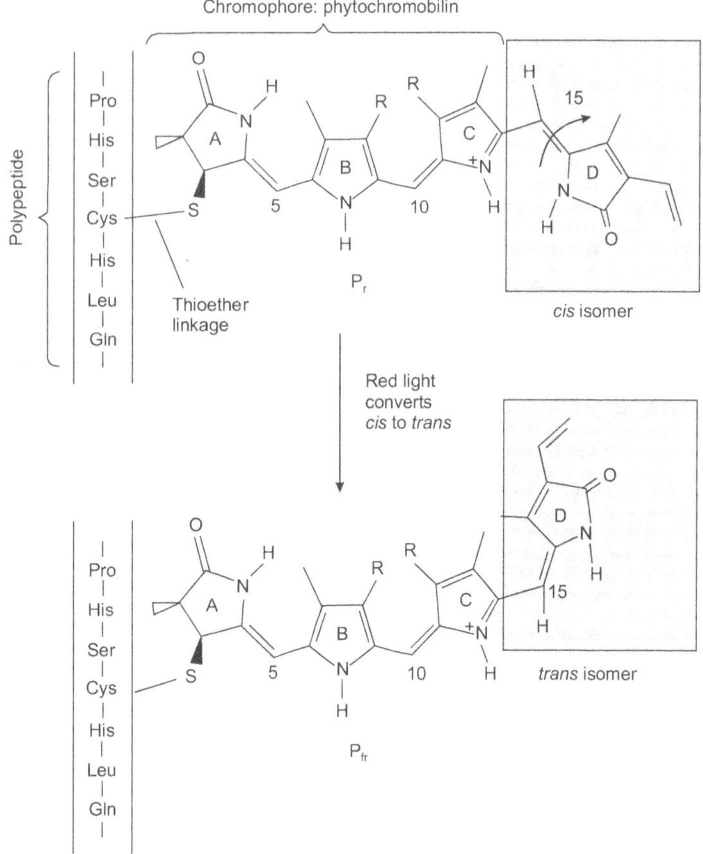

Figure 7.1   Structure of P$_r$ and P$_{fr}$ forms. The chromophore undergoes a *cis-trans* isomerization at carbon 15 in response to red and far-red light

Phytochromes can be classified into Type I and Type II phytochromes and are found to be encoded by a family of genes, named *phy*. There are five individual members—*phy*A, *phy*B, *phy*C, *phy*D and *phy*E.

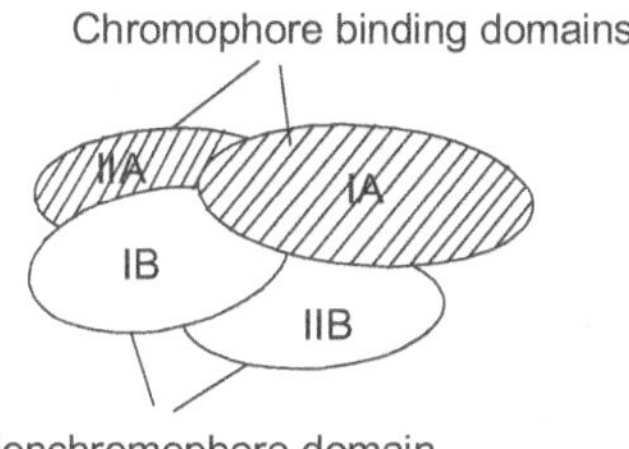

Figure 7.2 Structure of the phytochrome dimer

REGULATION OF *phy* GENES

phy gene family can be classified as either Type I or Type II based on their expression and products obtained. *phy*A is the only gene that codes a Type I phytochrome.

*phy*A gene is transcriptionally active in dark-grown, found to be present in high amount whereas *phy*B was found to be less expressed. The remaining *phy* genes (phyB → phyE) codes the Type II phytochromes.

The expression of *phy*A gene is strongly inhibited by the light in monocots. In dark-grown oat, treatment with red light reduces phytochrome synthesis because the P_{fr} form of phytochrome inhibits the expression of its own gene (Figure 7.3).

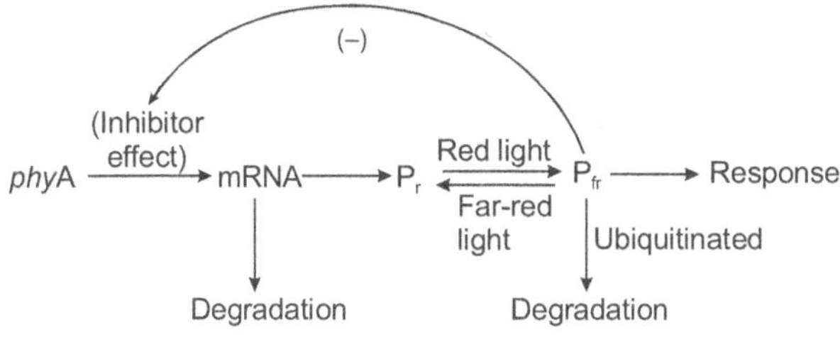

Figure 7.3 Expression of *phy*A on red light and its inhibitory effect

In addition, the PhyA mRNA is unstable, hence once etiolated seedlings are transferred to the light, PHYA mRNA rapidly disappears. The inhibitory effects of light on *Phy*A transcription in less in dicot, and in *Arabidopsis*, red light has no measurable effect on PHYA. PHYA levels also decline in the light as a result of proteolysis (ubiquitinated).

The remaining *phy* genes, *phy*B, C, D & E, encode the Type II phytochromes. Type I is abundant in dark-grown plants; in light-grown plants, the amounts of both types are about equal. The reason is that the expression of their mRNAs (PHYB–E) is not significantly changed by light and the encoded PHYB through PHYE proteins are more stable in the P_{fr} form than in P_{fr}A (Figure 7.4).

$$phyB\text{–}E \longrightarrow mRNA \longrightarrow P_r \underset{\text{Far-red}}{\overset{\text{Red}}{\rightleftharpoons}} P_{fr} \longrightarrow Response$$

Figure 7.4   Expression of *phy*B–E on light and there is no inhibitory effect

The creation of double and triple mutants of *Arabidopsis* was made to assess the relative role of each phytochrome in a given response. It was found that, like *phy*B, *phy*D plays a role in regulating leaf petiole elongation as well as in flowering time. Like analyses support the idea that *phy*E acts redundantly with *phy*B and *phy*D in those processes, but also acts with *phy*A and *phy*B in inhibition of internode elongation. *Phy*C is the least well characterized. *phy*C, D and *phy*E appear to play roles that are for the most part redundant with those of *phy*A and *phy*B. *phy*B appears to be involved in regulating all stages of development, whereas the functions of other phytochromes are restricted to specific developmental steps or responses. Based on many mutational studies, the structure of phytochrome holoprotein has been elucidated (Figure 7.5).

Phytochrome holoprotein have two domains: an N-terminal light-sensing domain which confers photosensory specificity to the molecule and a C-terminal domain that contains the signal-transmitting sequences. N-terminal domain consists of a site for chromophore binding and a site for photodegradation. C-terminal domain consists of a dimerization site, a ubiquitination site, a tag for degradation and a regulatory region. The C-terminal domain transmits signals to proteins that act downstream of the phytochrome.

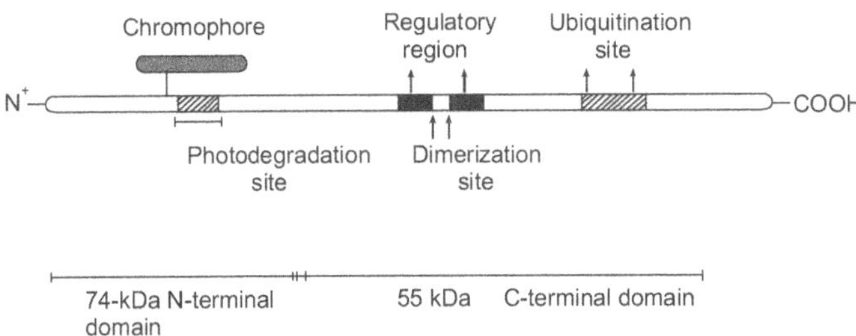

Figure 7.5 Schematic representation of the phytochrome holoprotein with various functional domains

PHYTOCHROME-CONTROLLED RESPONSES

Control in Vegetative vs Reproductive Development

1. Phytochromes control seed germination, leaf and stem growth, plastid development and flowering in higher plants. Flowering plants can be broadly classified into three types.

   1. short-day plants that flower in short days,

   2. long-day plants that flower in long days, and

   3. day-neutral plants which are insensitive to day length

Long-day plants require longer exposure to light in order to flower whereas short-day plants require shorter exposure to light in order to flower (Figure 7.6). In both the short-day and long-day plants, it is the length of the dark period that is important. If this dark period is interrupted with a flash of red light, their flowering response is altered. This flash of red light

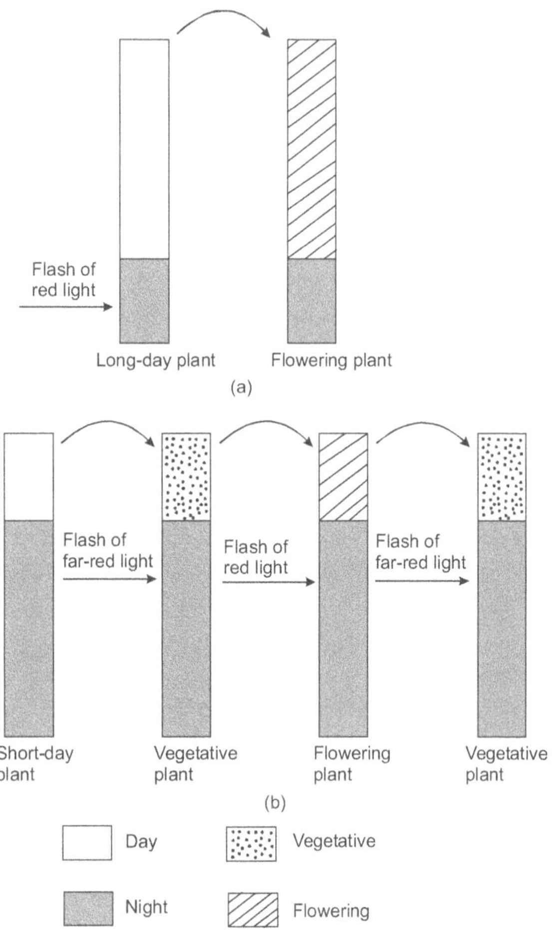

Figure 7.6 The effect of red (R) (660 nm) and far-red (FR) (730 nm) light on flower development in short-day and long-day plants

makes the long-day plants to start flowering and the short-day plants to remain vegetative. A single flash of far-red light on the other hand has no effect. However, the red light effect is reversible by far-red light, so that giving one red flash followed by one far-red flash makes the short-day plants to flower.

Similarly, the far-red effect is reversible by red light, so that giving 3 flashes in the sequence red, far-red, red results in vegetative growth of the short-day plant (Table 7.1).

Table 7.1   Response of short-day plants to flashes of red and far-red light

Flash of light during dark hours	Response in short-day plants
Red (660 nm)	Vegetative
Far-red (730 nm)	Flowers
Red flash followed by far-red	Flowers
Red flash followed by far-red flash again followed by red flash	Vegetative
Red flash followed by far-red flash followed by red flash followed by far-red	Flowers

2. Molecular mechanism of phytochrome-regulated changes in plants begin with absorption of light by the pigment. Absorption of light (primary signal or ligand) alters the molecular properties of phytochrome causing the signal-transmitting sequences in the C-terminus to interact with one or more components of a signal transduction pathway that ultimately bring about changes in the growth, development or position of an organ (Figure 7.7). Some of the signal-transmitting motifs appear to interact with multiple signal transduction pathways; others appear to be unique to a specific pathway. These responses fall into two general categories:

1. relatively rapid responses involving ion fluxes and

2. slower, long-term processes associated with photomorphogenesis, involving alterations in gene expression.

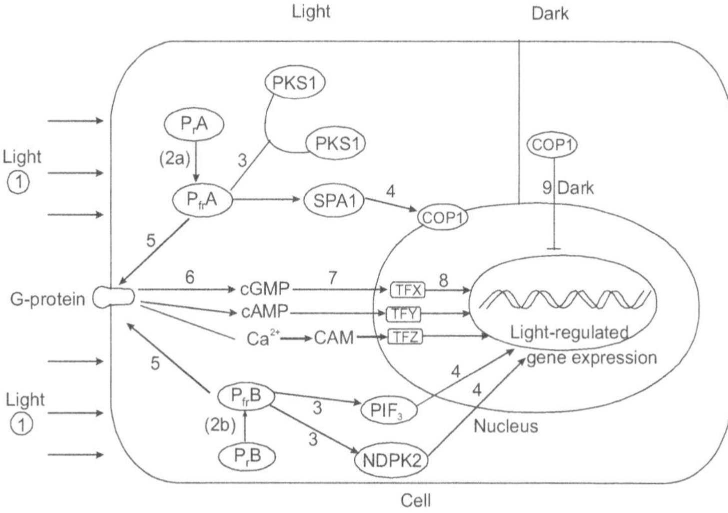

Figure 7.7 A diagrammatic representation of signal transduction by phytochrome (An overview) 1. Light source; 2a & 2b. Red light converts P_rA and P_rB to its P_{fr} forms; 3. Activated P_{fr} forms phosphorylate different kinases (PKS1, NDPK2) and other factors (PIF3 and SPA1); 4. Phosphorylated kinases and intermediate factors regulate the expression of light-responsive genes; 5. Activated P_{fr} forms of phytochrome activate membrane-bound G-proteins (receptors); 6. Activated G-proteins transduce the signal to secondary messengers (cGMP, cAMP and Ca^{2+}); 7. Secondary messengers activate various transcription factors (TFX, Y & Z); 8. Activated TF bind with TATA box of light-regulated genes along with kinases/factors that help in the initiation of respective genes; 9. In the absence of light P_r is not converted to the active P_{fr}, and COP1 proteins act as repressors for light-responsive genes.

As the term photomorphogenesis implies, plant development is profoundly influenced by light. Etiolation symptoms include spindly stems, small leaves and the absence of chlorophyll. Complete reversal of these symptoms by light involves major long-term alterations in metabolism that can be brought about only by changes in gene expression. The stimulation and repression of transcription by light is very rapid, say, within 5 minutes. These early-gene expression is regulated by the direct activation of transcription factors by one or more phytochrome-initiated signal transduction pathways. These transcription factors enter the nucleus and stimulate the transcription of specific genes. These genes are said to be late genes and are the small subunits of ribulose1,6-bisphosphate carboxylase/oxygenase (RuBisCO) and the major light-harvesting chlorophyll *a/b*-binding proteins (Refer chapter 2) associated with the light-harvesting complex of photosystem II (LHCIIb proteins). The proteins of these genes RBCS and LHCB (also called CAB proteins) play important roles in chloroplast development and greening and these genes are present in multiple copies in the genome.

PHYTOCHROME MUTANTS

The photomorphogenic mutations have been studied at 13 points in *Arabidopsis* and some are shown in Figure 7.8.

Among these mutations, *cop* and *det* are interesting because the wild type products of these genes act to repress the de-etiolation of seedlings. Hence, mutations in these genes arrest the repressing activity and encourage de-etiolation. Among the various *cop* genes, *cop* 1 has been studied in detail. A mechanism for the light control of *cop* 1 gene repression is suggested by the studies on expression of a fusion protein in onion cells. The fusion protein construct consisted of the reporter protein GUS fused to COP 1, which allowed the location of COP 1 to be monitored by histochemical staining for GUS. In the dark, COP 1 is located in the nucleus whereas in the light it moved into the cytoplasm. The results were remarkable and thus it is clear that light affects the location of this protein, COP 1.

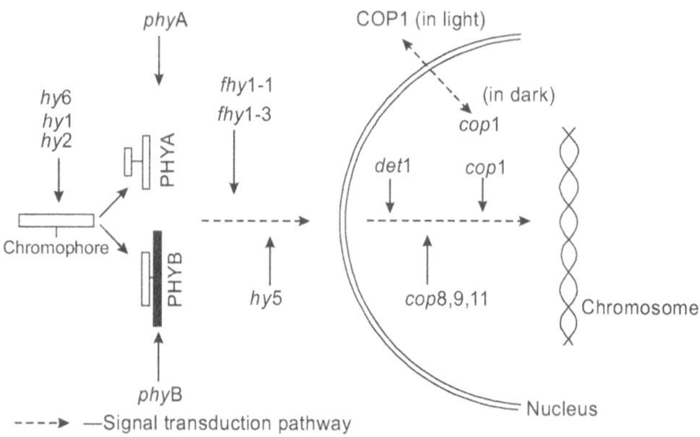

Figure 7.8 Some of the photomorphogenic mutations in *Arabidopsis*. Genes indicated in italics represent the points of mutations. These mutations are thought to have an effect on light reception and signal transduction.

When COP1 is transported into the nucleus in the dark, it suppresses the photomorphogenetic genes whereas in the presence of light, it exits the nucleus and allows photomorphogenesis to take place. Thus, COP1 protein functions as a transcriptional repressor along with other proteins. Further studies indicate that COP and DET proteins act at the downstream of signal transduction pathway involved in phytochrome action (Table 7.2).

Table 7.2 Mutation in phytochromes and its effects

Genes where mutations occur	Effects observed
phy6, phy1, phy2	Photoreceptors PHYA, and PHYB production is affected
fhy1-1, fhy1-3, hy5	Signal transduction pathway impaired
cop, det	De-etiolation repressed

CONCLUSION

Light has acted as a signal to induce a change in the form of the seedling, from one that facilitates growth beneath the soil to one that is more adaptive to growth above ground. Phytochromes play an important role in the growth and development of a plant.

REVIEW QUESTIONS

1. What is photomorphogenesis and what is its role in plant development?

2. What are the genes encoding phytochrome proteins?

3. Describe how phytochrome controls gene expression in plants.

SEED STORAGE PROTEINS

INTRODUCTION

Of the edible proteins produced in the world, 70% comes from the seeds and therefore these proteins are of considerable economic importance. Seed proteins can be divided into the following two classes.

1. **Storage proteins**—proteins that are unique to the seed.

2. **Metabolic proteins**—proteins that are also present in the rest of the plant.

The storage proteins are synthesized in the developing seed and stored for use during germination. In legumes, the major components of seed storage proteins are globulins and albumins, whereas in cereals, the major components are prolamins and glutelins. The essential amino acids in the seed storage proteins are leucine, isoleucine, lysine, methionine, phenylalanine, threonine, tryptophan, tyrosine and valine. The globulins in legume seeds account for 20–25% of the seed dry weight and the prolamins and glutelins in cereals account for 8–15% of the seed dry weight. These proteins have unbalanced amino acid content for mammalian diets, as they are deficient in lysine, threonine and tryptophan.

In view of their importance as food, seed storage proteins have been the subject matter of a great deal of molecular investigation. Molecular studies have therefore been able to provide the fundamental information about the proteins themselves, as well as the genes that encode them or control their synthesis. One aspect of storage proteins that has received considerable attention is the possibility of altering the amino acid composition of these proteins in order to improve their nutritional quality.

Seed storage proteins are synthesized as pre-proteins with an N-terminal signal peptide that causes them to enter the endoplasmic reticulum. Within this, they are modified before being deposited in membrane-bound protein bodies derived from the ER or the vacuole.

The proteins accumulate on the vacuolar side of the tonoplast (the membrane which surrounds the vacuole), and as they accumulate, the tonoplast appears to evaginate around the protein deposits to pinch off protein bodies. The vacuole gradually gets smaller as more and more protein is deposited and more protein bodies are formed. These protein bodies also contain a number of hydrolytic enzymes that are commonly found in vacuoles. The site of protein accumulation (the vacuole) and the site of protein synthesis (the RER) are spatially separated in the cell and the transport of the proteins within the cell involves the Golgi apparatus. There are direct tubular connections between the RER and the cisternae of the Golgi apparatus.

In legume seeds, the storage proteins accumulate predominantly in swollen cotyledons of the embryo, whereas in cereals they are accumulated mainly in the endosperm. In legumes which fix atmospheric nitrogen, 20–40% of the seed dry weight is protein, whereas in cereals it is from 7–16%.

Among the various storage proteins, the **zein** (prolamin) storage protein of maize and the **legumin** and **vicilin** (globulins) storage proteins of pea are studied in depth.

ZEIN PROTEINS OF MAIZE (*Zea mays*)

Zein proteins are synthesized by membrane-bound polyribosomes in the cytoplasm of developing maize (monocotyledon) endosperm (Figure 8.1).

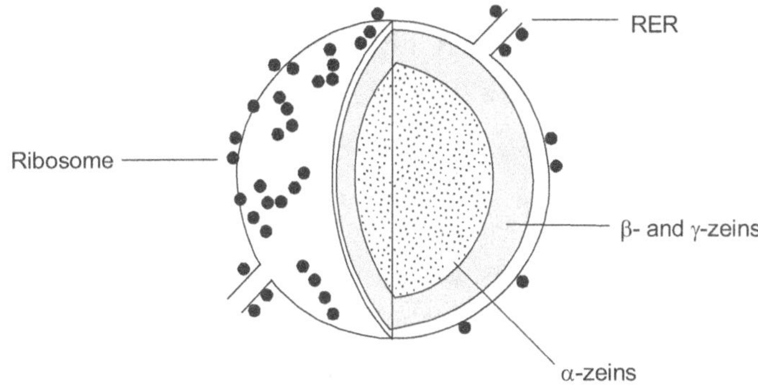

Figure 8.1   Structure of maize endosperm protein body containing zein storage proteins

These proteins are co-translationally transported into the lumen of the rough endoplasmic reticulum (RER) where they aggregate into insoluble masses called **protein bodies.** Zein synthesis occurs during the period between free nuclear division within the endosperm and desiccation of the seed. At sixteen days after pollination, it has been calculated that in each endosperm, zein proteins are synthesized at 300 mg/day. Zein proteins accumulate during the period from 10 to 28 days after pollination. The following are a few examples of zein proteins.

Molecular weight (kDa)	Type of zein protein
27	γ-zein S-rich
22	α-zein
19	α-zein
16	γ-zein S-rich
14	β-zein S-rich
10	δ-zein S-rich

S-rich—Sulphur rich

The α-, β- and γ-zeins differ significantly in their structure and location in the protein body.

Genes Encoding α-Zeins

A number of α-zein genes have been cloned, both as cDNA and as genomic DNA. Analysis of these clones revealed the structure of proteins and the following details.

There are 20 structural genes for α-zeins. The genes are distributed as follows:

Chromosome	Genes
Short arm of chromosome 4	8
Long arm of chromosome 4	2
Short arm of chromosome 7	9
Long arm of chromosome 10	1

Characteristics of α–zeins

1. The two α-zeins of different molecular weights (22-kDa and 19-kDa) are structurally related.

2. The 22-kDa α-zeins are encoded by a multigene family with about 25 members, which have 90% homology among them.

3. The 19-kDa α-zeins are encoded by a family of about 55 members.

4. The genes contain no introns.

5. The α-zein amino acid sequence predicted from the cDNA sequence is larger than that calculated from the SDS-PAGE. This is because the primary polypeptide includes an N-terminal 21-amino acid signal sequence that is responsible for transport of the protein across the endoplasmic reticulum (ER) to the lumen, where the accumulated proteins form protein bodies. This signal peptide is removed during transport resulting in a 5-kDa reduction in the size of the mature protein.

6. Both the α-zeins contain a central domain with a 20-amino acid sequence that is tandemly repeated. The 20-amino-acid repeats form α-helical secondary structures that in turn stack antiparallel to form tertiary structure. This structure is important in the preservation of the protein during dehydration. During rehydration and seed germination, the protein bodies are degraded by proteases.

Genes Encoding β- and γ-zeins

The β- and γ-zeins are each thought to be encoded by a single gene.

Genes Controlling Zein Synthesis in Maize

In addition to the structural genes for zein proteins, a number of genes have been identified that affect the synthesis of zein proteins. Some of them are listed in Table 8.1.

Table 8.1   Genes affecting zein synthesis

Gene	Chromosome location	Dominance	% inhibition of zein synthesis	Zein protein affected
opaque-2	7 (short)	Recessive	47	22 kDa
opaque-6	?	Recessive	88	All
opaque-7	10 (long)	Recessive	77	19 kDa
floury-2	4 (short)	Semi-dominant	34	All
Defective-endosperm-B30	7	Dominant	12	22 kDa

These genes are present on chromosomes 4, 7 and 10. However, although there is linkage between some of the zein structural genes and these controlling genes, the control function is not limited to the linked structural gene. Thus these genes transcribe some *trans*-acting regulatory factors.

Table 8.1 illustrates the complexity of the control system for these genes because it shows that mutations in a number of genes can affect zein synthesis. Some of the mutations are recessive (*opaque-2*) while others are semi-dominant or dominant.

The effect of some of the genes may be either additive or epistatic, i.e., prevent the expression of other similar genes or are lethal. Therefore, these genes may not be specific for zein protein synthesis but may be controlling the expression of other genes also. *opaque-2* has been well characterized. The wild type allele of *opaque-2* produces a 2-kb transcript encoding a 51-kDa polypeptide. *opaque-2* mutants produce no *opaque-2* mRNA. Analysis of the predicted *opaque-2* amino acid sequence shows that the protein contains a leucine zipper and zinc finger DNA-binding domains. The sequence has homology with the yeast transcription factor, GCN4, and it has been shown that the wild type *opaque*-2 gene can complement a yeast *gcn*4 mutant. This provides an experimental confirmation that *opaque*-2 encodes a transcription factor that is a *trans*-acting regulatory factor.

LEGUMINS AND VICILIN/CONVICILINS IN PEA (*PISUM SATIVUM*)

The dicotyledon garden pea (*Pisum sativum*) consists of globulins and albumins as its seed storage proteins, with the globulins (60%) being the major component. The globulins can be separated into two groups on the basis of the size of the aggregated polypeptides. The **legumins** have a sedimentation coefficient of 11–12S and the **vicilins and convicilins** have a sedimentation coefficient of 7S.

The globulins are synthesized in the storage parenchymal cells of the cotyledons rather than the endosperm as in maize. Similar to maize, synthesis occurs during the later stages of seed development. Like maize, the proteins are synthesized in the cytoplasm by ribosomes bound to the ER and co-translationally transported into the lumen of the RER (rough endoplasmic reticulum). This transport is also accompanied by the removal of an N-terminal signal peptide.

Legumins and vicilin/convicilins are found in the protein bodies in pea cotyledons but the protein bodies do not arise simply as protein accumulations within the RER.

Legumin Structure

Information from cDNA clones, genomic clones and protein structure indicates that there are 4 classes of pea legumin.

They exist as dimers of structurally different polypeptide subunits (α and β) that are bonded together via a disulphide bridge. The heterodimer is synthesized as a single 60-kDa precursor, with the structure NH_2-α-β-COOH. The covalent peptide bond between the α and β polypeptides is thought to be cut after the proteins have been deposited in the protein bodies. The legumin dimers aggregate to form a large oligomeric structure of 362 to 400 kDa with a sedimentation coefficient of 11–12S.

Vicilin/Convicilin Structure

The vicilins are heterogeneous group of polypeptides with a molecular weight of 12–17 kDa. There are five classes of vicilins.

The heterogeneity arises from the following factors:

i. Existence of gene families

ii. Proteolytic processing after translation

iii. Post-translational glycosylation of polypeptides

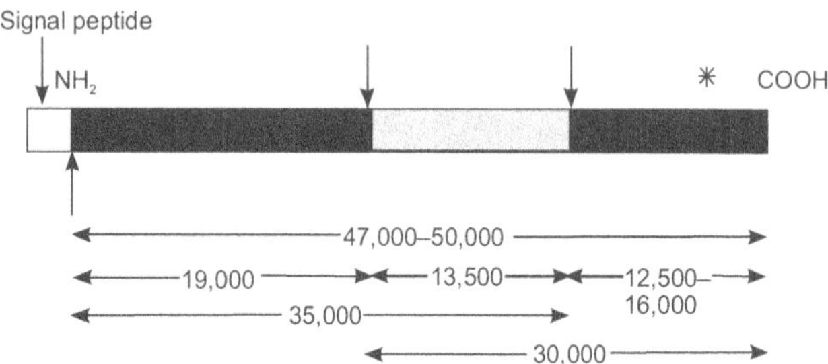

Figure 8.2　Pattern of post-translational processing of the 47–50-kDa pea vicilin precursor polypeptide to give rise to many polypeptides. Asterisk indicates the glycosylation site; ↓ indicate the proteolytic processing site.

The vicilins are synthesized as precursor polypeptides of 47–50 kDa. Figure 8.2 represents the post-translational processing of the 47–50-kDa vicil in mature polypeptides.

The molecular weight of the C-terminal fragment depends on its glycosylation. Convicilin is equivalent to the 50-kDa vicilin polypeptide with a 120–166 sequence of hydrophilic amino acids inserted near the N-terminus.

Pea Globulin Genes

The number of legumin genes in the pea genome has been estimated to be more than 10. The structural genes for legumin have been located at 3 loci on chromosome 7. The legumin *legA* gene contains 3 small introns. There are approximately 24 vicilin genes grouped into 7 loci on chromosome 7 and there are 2 convicilin genes on chromosome 2.

In order to analyse the region of the promoter that is essential for the expression of legumin gene, deletions were made in the full-length legumin gene promoter. These truncated promoters were fused to the legumin coding region and transformed into tobacco plants. Plants were then screened for the production of the protein, legumin, so as to identify the minimal length of the promoter required for expression.

The structure of the genomic clone of the pea *legA* gene is shown in Figure 8.3. The ATG translation start site is shown together with the exons (solid bars) and the polyadenylation signal (Figure 8.3a). The region upstream of the ATG site has been expanded (Figure 8.3b) to show the following sequence features:

1. TATA and CAAT boxes proximal to the translation start ATG

2. a 28-bp "legumin box" sequence, found in the putative promoter regions of a number of legumin genes

3. sequences similar to cereal glutenin gene control sequences which are situated distal to the translation start ATG

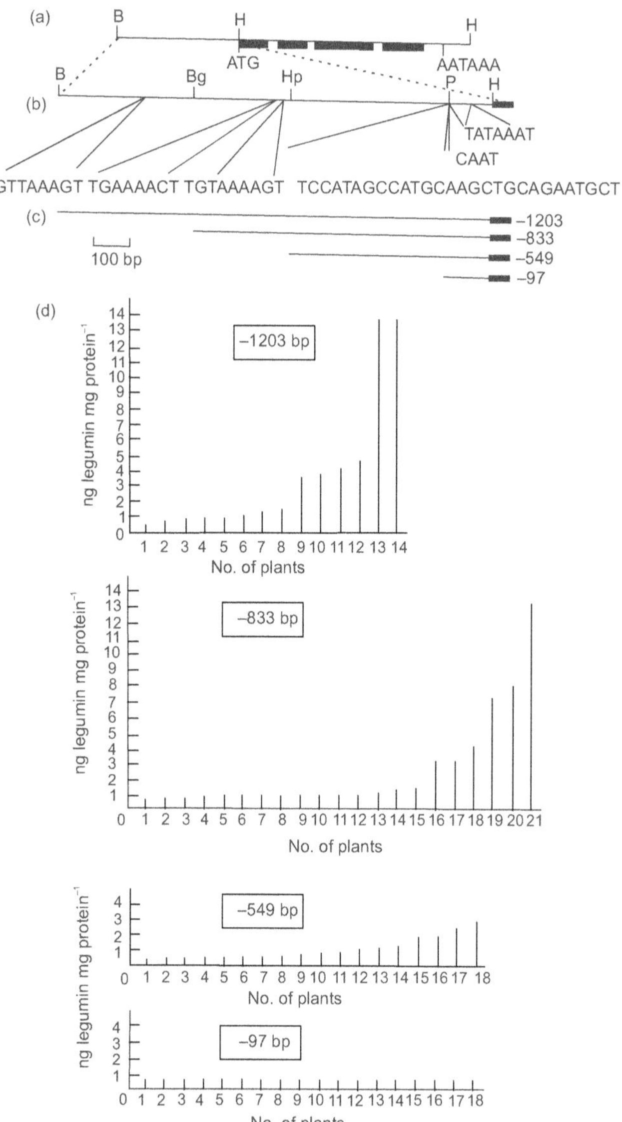

Figure 8.3 Analysis of the promoter controlling legumin gene expression

The 5'-region of the legumin gene contains –1203 bp upstream of the transcription start site. A series of deletions were made that reduced the length of this putative promoter. The deletion constructions are identified by the number of bases they contain (Figure 8.3c).

- 1203 contains all putative control genes

- 833 has one of the "glutenin control" sequences deleted

- 549 has all three of the "glutenin control" sequences deleted

- 97 has the legumin box plus all three "glutenin control" sequences deleted.

Transgenic tobacco plants were produced containing one of this deletion series of promoters plus the legumin gene. Legumin gene expression in transgenic seeds was measured by quantifying the production of the pea protein, using a legumin-specific antibody. The results are shown in Figure 8.3d where the levels of legumin are given for a number of transgenic plants containing each promoter construct.

As indicated in the Figure 8.3d, the minimal –97 bp promoter sequence which contains only the TATA and the CAAT boxes is not sufficient to provide seed expression of the legumin gene.

Increasing the promoter sequence to –549 bp that includes the legumin box, confers a low level of seed expression. However, high levels of expression were seen only in plants that contained the legumin transgene plus over –549 bp of 5' flanking sequence.

CONCLUSION

Several conserved DNA sequences have been found in the upstream regions of genes encoding seed storage proteins (Kreis *et al.*, 1985) including a vicilin box and a legumin box. One or more of these sequences may be involved in regulating the expression of these genes. These signals resemble the viral and animal 'enhancer' core sequences, which stimulate gene expression. No direct test of their function in plants has yet been made. However, recent experiments suggest they may have regulatory significance. Further work is necessary in order to elucidate the role of this sequence and the protein factor in regulating expression of the zein genes.

REVIEW QUESTIONS

1. What are the genes encoding zeins in maize and the factors controlling their expression?

2. Give an account of the genes involved in the synthesis of zein storage proteins.

3. Write short notes on seed storage proteins.

9

INDUCIBLE CONTROL OF GENE EXPRESSION

INTRODUCTION

The "Green revolution" led by Norman Borlaug, Monkombu Swaminathan and Gurdov Khush enabled the world's food supply to be tripled during the last three decades of the 20th century. An extraordinary increase in agricultural productivity was made possible by the adoption of genetically improved varieties coupled with advances in crop management. The world's population has increased at an alarming rate with nearly 800 million people going to bed hungry every day. So, there is a need for more food production in an environmentally friendly way. Countries inhabited by half the world's population are already experiencing water crisis, while the high agrochemical inputs that maximize yields exert a high environmental impact, which is not acceptable. There is a desperate need to produce more food from less land with less water and reduced agrochemical inputs.

The majority of agricultural scientists are convinced that the required high-yield, high-quality, low-cost, low environmental-impact crops can be delivered by the exploitation of the techniques for plant biotechnology in molecular breeding strategies. There is an increasing interest in the use of inducible systems for the expression of novel genes introduced into plants.

For this, a wide range of promoters of different origins is used. The promoters are broadly classified into 3 groups.

1. Promoters introduced from non-plant backgrounds such as bacteria (*Tn10*), insects and animals (ecdysteroid and glucocorticoid) and chemical inducers (copper).

2. Promoters responding to particular environmental signals such as nitrate, heat-shock and wounds.

3. Promoters regulated by native plant hormones such as cytokinins, abscisic acid and auxins.

Class I promoter systems utilize transcriptional control systems from outside the plant genome. These systems require external inducers, for example, tetracycline for *Tn10*-encoded tetracycline repressor system, synthetic hormones and agonists for non-plant hormone-responsive systems and copper for copper-controllable promoter systems.

Class II promoter systems make use of environmental signals. As plants survive in a rigorous environment, they have developed a wide range of mechanisms for defence against diseases, insect or other predator attack and are able to respond to a wide range of chemical or nutritional threats provided by the environment. These mechanisms include wound inducibility, nutrient control of expression (nitrate inducibility), and the heat-shock response. Thus, class II differs from class I with the absence of external chemical or synthetic inducers.

Recent research has described specific sequence elements in hormone-responsive promoters, the class III promoters which will allow the controlled expression of characters by endogenous regulatory pathways. Plant developmental processes such as organogenesis, flowing, fruiting and abscission offer characterized promoter systems that are amenable for the controlled expression of introduced genes. The utility of inducible promoter systems in laboratory-based research is self-evident and there is clear potential for their usefulness in the genetically modified plants of the future.

CLASS I PROMOTERS

Tn 10-ENCODED TETRACYCLINE REPRESSOR

Tet repressor (TetR) is a protein molecule that regulates the expression of its own gene (*tetR*) as well as the expression of the tetracycline-resistant (tc-resistant) gene (*tetA*) in gram-negative bacteria.

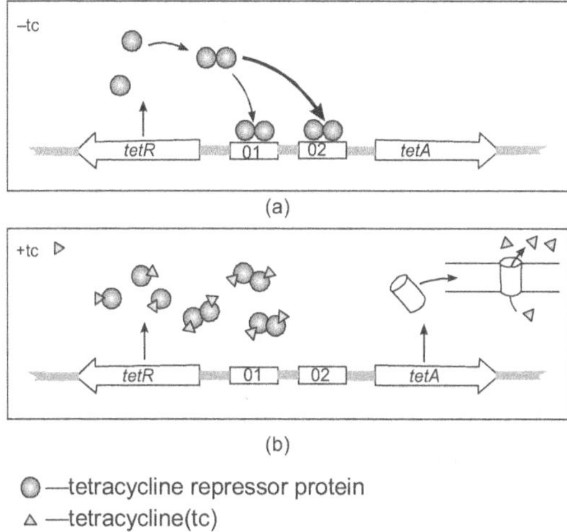

Figure 9.1 Genetic organization and mechanism of regulation of the *Tn10*-encoded tc-resistance determinant. (a) In the absence of an inducer (b) In the presence of an inducer

The genes *tetR* and *tetA* are oriented with divergent polarity with a central regulatory system consisting of overlapping two *tetR* operators with two *tetA* promoters (Figure 9.1). In the absence of tetracycline (tc), *tetR* dimers bind to the tet operators (constitutively expressed) thereby resulting in the repression of both genes. Thus the expression of *tetA* is suppressed in the absence of tetracycline (tc). In the presence of the inducer tetracycline (tc), a TetR–tc complex is formed. Binding of tc to TetR leads to a conformational change in TetR rendering a non-DNA-binding conformation to the protein and preventing the binding of the complex to the operators. Thus, the promoters of *tetA* are free for the recruitment of transcription factors (TF) and RNA polymerase resulting in the expression of *tetA*. Thus induction is mediated by tetracycline (tc), which binds to TetR, thereby removing its DNA-binding activity. When there is no repression by TetR, *tetA* is expressed and the product is an integral membrane protein that pumps tc out of the cell, thus making the organism tetracycline-resistant. Dissociation of the bound TetR from the DNA leads to transcription of both genes.

Taking advantage of the high specificity of the TetR-tet operator interaction, tc-regulatable gene expression systems have been developed for a variety of eukaryotes.

TetR as a Repressor

In prokaryotes, the TetR repressor protein prevents transcription by binding to the tet operator sequence. This principle of regulating gene expression has been applied to control a plant promoter. Two of these *tetR* operator sequences were positioned flanking the TATA box of the cauliflower mosaic virus (CaMV) 35S promoter. The *tetR* gene was placed upstream the CaMV promoter. Now the constitutively expressed TetR (repressor protein) on binding to the two cloned *tetR* operators suppressed the expression of the CaMV promoter. Repression was relieved by the addition of tc (Figure 9.2).

When this module was used in plants, the transgenic plants did not efficiently express the system. Hence two adjustments were made.

1. The untranslated leader sequence was reduced by 50 bp bringing the enhancer module closer to TATA box and this induced the formation of one million TetR molecules per cell.

2. The location of the operators was adjusted.

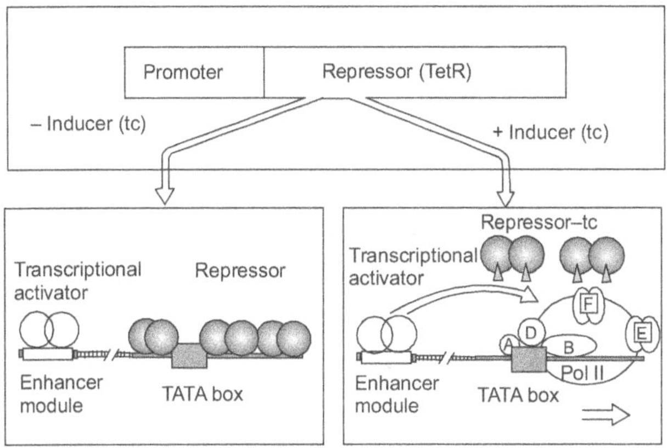

Figure 9.2   Diagrammatic representation of the use of TetR to repress transcription

Integration of 3 operators into the CaMV 35S promoter led to the development of a tightly repressible CaMV promoter.

In a prokaryotic system, the TetR protein prevents binding of a single protein, RNA polymerase, whereas in a eukaryotic system, TetR has to interact with 40 proteins, which cooperate to form a functional transcription initiation complex (Roeder, 1991).

In the absence of the inducer tc, binding of TetR to the operators prevents the assembly of the transcription initiation complex at TATA box (Figure 9.2). In the presence of tc, tc binds to TetR, which triggers a conformational change in the protein, so that it no longer binds to DNA. Thus, the multifactorial initiation complex, which contains TFIID, TFIIA, TFIIB, TFIIF and TFIIE, other associated factors and RNA polymerase II, can assemble and transcription is initiated.

Induction of gene expression was achieved with as low as 0.1 mg tetracycline/L. Under these conditions, induction was extremely fast (10 minutes) implying that the signal transduction chain is very short.

TetR as an Activator

We have already encountered an example of a repressor (the TetR repressor protein) in the previous section. Whereas a repressor is a protein that prevents transcription on binding to an operator sequence, now we introduce the term activator, which refers to a protein, which initiates transcription on binding to an operator.

TetR can be turned into a tc-controlled transcriptional activator (**tTA**) when fused to the potent transcriptional activation domain of herpes simplex virus protein 16 (VP16). This addition results in the C-terminal extension of TetR. Although the size of TetR increases because of this addition, it retains its DNA-binding activity and tc inducibility.

A target promoter containing 7 *tet* operators upstream a minimal promoter was used to check the efficiency of the tTA as an activator. Fifty per cent occupancy of binding sites by tTA was found to be sufficient for transcriptional activation.

The fusion protein consisting of TetR and an activation domain (tTA) is positioned under the control of a strong constitutive promoter, and it controls a target promoter in a tc-dependent manner (Figure 9.3). Multimerized operators are indicated as a black box in brackets. In the absence of the effector tc (filled triangles), binding of tTA (grey pear-shaped) to the operator activates transcription, by favouring the functional assembly of the initiation complex consisting of TFIID, TFIIA, TFIIB, TFIIF, TFIIE, other associated factors and polymerase II.

Binding of tc to tTA triggers a conformational change in the protein, so that it can no longer bind to DNA and transcription is activated. Tissue specificity of the system can be achieved by choosing appropriate promoters to drive expression of tTA.

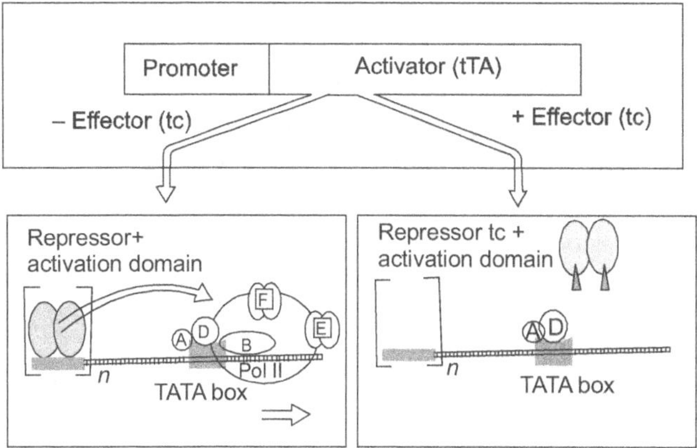

Figure 9.3   Diagrammatic representation of the use of TetR to activate transcription

Thus the promoter system can be shut off in the presence of tc. In other words, any target promoter can be inactivated by using tetracycline. This system was shown to work in tobacco plants (Weinmann *et al.*, 1994).

HORMONE–INDUCIBLE GENE EXPRESSION

Inducible regulation of transgenes in plants has been achieved by relief of repression or activation of transcription. In the case of tetracycline system (bacterial origin-tetR), operator sequences were inserted within the target promoter region. The repressor protein binds to these operator sequences in the absence of the ligand (tetracycline), preventing transcription of the target gene. Addition of the ligand prevents repressor binding to the operator sequences, and thus transcription is initiated.

Due to the nature of their inducing chemicals (tc), it is likely that these systems will be restricted to research applications only. Although the utility of these systems for research purposes has been documented, a number of them are not tightly regulated and exhibit low levels of inducible expression or utilize chemistry which is phytotoxic or incompatible with agricultural use.

To overcome this problem, animal hormone receptor/ activator are used in gene transfer in which the ligands are their respective hormones. Since these hormones are of animal origin, certain domains were made of use in the construction of transcriptional factor (TF) or activator of repressor. Two systems have been discussed here: GVG and ERS systems.

In order to understand hormone-inducible gene expression, a detailed knowledge about hormonal receptors and nuclear receptors otherwise called transcription factors are necessary.

Hormone Receptors

Two groups of hormone receptors are present in cells.

1. Intracellular receptors—for molecules that can enter the cell (Figure 9.4).

2. Plasma membrane receptors—for molecules that cannot enter the cells.

The topic of interest here is the intracellular receptors, i.e., receptors for hormones that are able to enter the cell. These are ligand-activated transcription factors that are localized in the nucleus or cytoplasm. Hormones that bind to these intracellular receptors are those that can cross the membrane usually. Examples of these hormones are small hydrophobic molecules or lipophilic molecules such as steroid hormones, thyroid hormones, 1,2,5-dihydroxycholecalciferol, and retinoic acid (Figure 9.5).

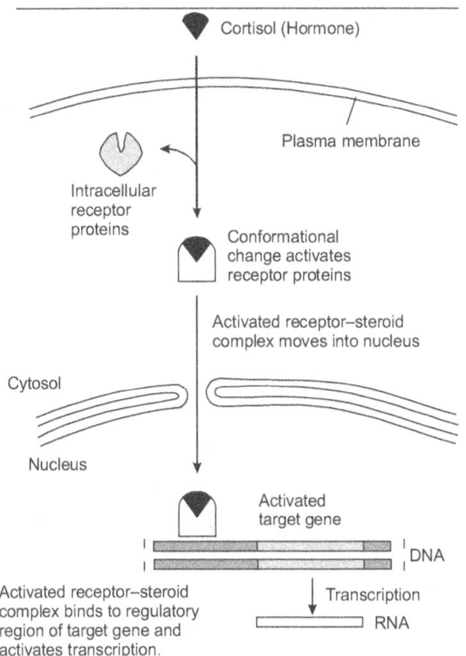

Figure 9.4   Hormone-induced DNA transcription

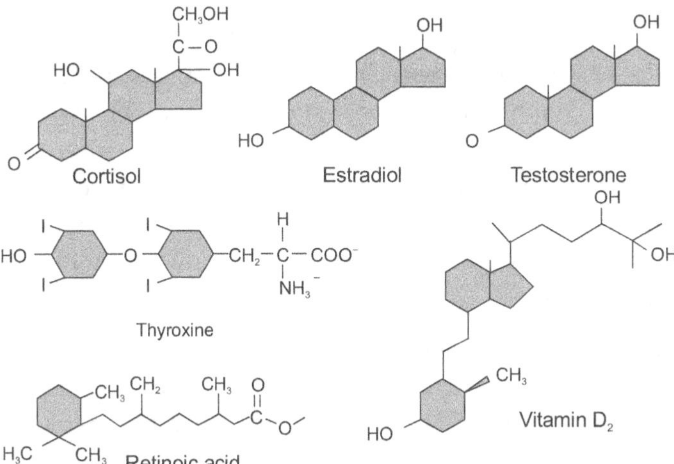

Figure 9.5   Ligands that bind intracellular receptors

Nuclear Receptors

Nuclear receptors are a large well-defined family of transcription factors with over 150 members. They have 6 different protein domains, designated as A, B, C, D, E and F. Each has a characteristic function (Figure 9.6).

Domains A and B are involved in ligand-independent transactivation. Domain C is a DNA-binding domain with the conserved region within the superfamily. It varies with 66 to 68 amino acids and has 8 invariant cysteine residues implicated in the formation of zinc fingers which are responsible for interacting with DNA responsive elements. It also contributes to dimer formation. Domain D is a hinge region, is variable in size and contains the sequences required for direct targeting to the nucleus. Domain E is the ligand-binding domain also known as the hormone binding domain (HBD) which is well-conserved. It is not only involved in interaction with ligands but also in the dimerization and ligand-dependent activation. Domain F is highly variable in size.

The nuclear receptor superfamily comprises four classes of receptors.

Class I : Steroid receptors   Steroid receptors are homodimers normally found bound by heat-shock proteins (HSPs) (70/90) and p^{59} to form a complex in the cytoplasm, e.g. GR—Glucocorticoid receptor, when bound by ligand, the receptor is released from the complex allowing translocation into the nucleus and binding to a cognate response element as a dimer. These are found only in vertebrates.

Class II : Retinoic-X-receptors (RXR)   These are heterodimers, belonging to the retinoic/thyroid receptor family. These receptors may interact with response element as heterodimers. They are normally found bound to DNA in the absence of ligand. An example is the EcR—ecdysteriod receptor of insects which interacts with ultraspiracle (USP: insect homologue of RXR) to form a heterodimer responsive to ecdysone.

Class III: Dimeric orphan receptors   These receptors bind as homodimers to direct repeats in the target promoter, e.g. RXR or COUP (chicken oralbumin upshroam promo TF-1).

Class IV: Monomeric orphan receptors   They are similar to class III receptors but bind to promoters as monomers, e.g. SF-1.

GLUCOCORTICOID INDUCIBLE GENE EXPRESSION IN PLANTS

Many genes have been introduced into plants and expressed under the control of various promoters. Many native promoters characterized in the literature can be used to express a transgene. Promoters active in specific cell types and responding to specific environmental stimuli or plant hormones are useful in limiting the location or the timing of transgene expression. Transgenic experiments with well-characterized promoters have given valuable evidence for functions of plant genes.

Regulatory Mechanisms of the Glucocorticoid Receptor

The glucocorticoid receptor (GR) is a member of the nuclear receptor superfamily. This family includes over 150 transcription factors that are receptors for various hydrophobic ligands including thyroid hormones, vitamin D, retinoic acid and steroids.

Each nuclear receptor functions as a sensor for the expression of its target genes by binding to specific *cis*-acting sequences. The domains of nuclear receptors have already been discussed.

Nuclear receptors consist of at least 4 domains (Figure 9.6a).

1. A and B—*trans*-activating domains
2. C—DNA binding domain
3. D—Hinge domain
4. E—Hormone-binding domain (HBD)

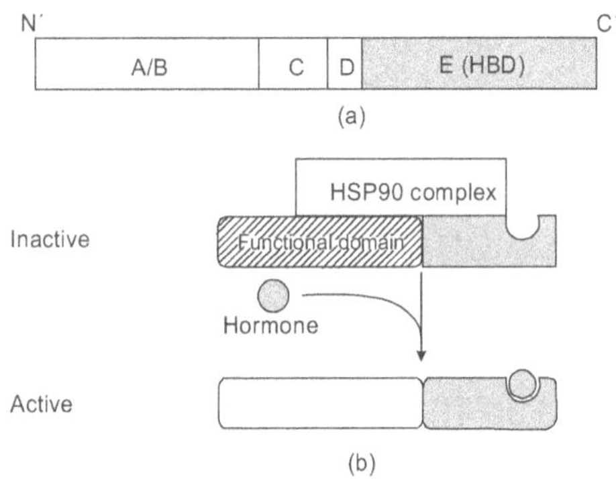

Figure 9.6   (a) General structure of nuclear receptor proteins. (b) Model for the regulatory mechanism of HBDs (Picard, 1994).

Many systems for the artificial regulation of gene expression have been developed, which use nuclear receptors and their specific *cis*-acting elements.

In the absence of ligands (hormones), HBDs repress the function of neighbouring domains by forming a complex with multiple proteins including the heat-shock protein HSP90 (Figure 9.6b). Ligand binding releases the complex resulting in derepression (expression). The functions of a protein can be artificially controlled by fusing HBD of glucocorticoid with a steroid hormone receptor.

Construction of the GVG System

HBD of mammalian glucocorticoid receptor functions in transgenic plants. Hence a transcriptional induction system was constructed using the HBD as a regulatory domain in a chimeric transcription factor consisting of the DNA binding domain of the yeast GAL4 (*gal*4 is a gene which belongs to the yeast galactose operon), the transactivating domain of the herpes viral protein VP16 and the HBD of rat glucocorticoid receptor (Figure 9.7). The resulting transcription factor was designated as GVG because it consisted of one domain each from **GAL4, VP16** and **GR**.

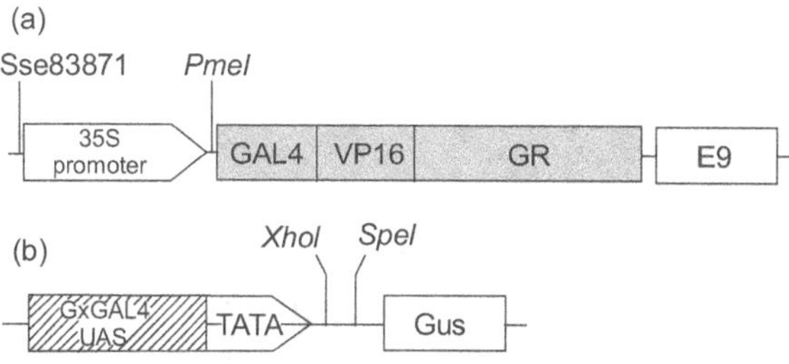

Figure 9.7 Structures of the *trans-* and *cis-*constructs in the GVG system. (a) Structure of the *trans-*construct. (b) Structure of the *cis-*construct Gus-β-glucuronidase, reporter gene

The GAL4 protein expressed from the *trans*-construct (Figure 9.7a) binds to specific DNA sequences designated as GAL4 upstream activating sequences (UAS) in the *cis*-construct (Figure 9.7b). The minimum DNA-binding domain sequence of GAL4 gene sequence (amino acids 1–74) was used in the *trans*-construct.

The VP16 (herpes viral protein) domain is a strong transactivator interacting directly with general transcription factors. Amino acids (413–490 of VP16) were fused to the C-terminus of GAL4 domain. The resulting fusion protein, GAL4–VP16, strongly activated the transcription from a promoter containing six copies of UAS in tobacco plants.

HBD of the rat GR (amino acids 519–795) was added to this strong transcription factor. The GVG protein strongly activates transcription from a promoter containing UAS only in the presence of glucocorticoid.

Induction Experiments with the GVG System

A major requirement for an ideal induction system is that the inducer must not evoke pleiotropic effects in plants. Another requirement is that the non-induction level of transgene expression should be minimal or absent.

The inducibility of the GVG system has been studied in transgenic tobacco plants (Aoyama and Chua, 1997). The 35S-driven *gvg* gene was introduced into transgenic tobacco together with a *cis*-construct containing the luciferase (*luc*) gene as a reporter. Induction of luciferase activity was observed when the transgenic tobacco plants were treated with dexamethasone (DEX), a strong synthetic glucocorticoid. The maximum expression level was over 100 times that of non-induction levels. The induction levels correlated with the DEX concentration ranging from 0.1 to 10 mM, when the plants were grown on an agar medium containing DEX.

ECDYSTEROID–AGONIST INDUCIBLE CONTROL OF GENE EXPRESSION IN PLANTS

Ecdysone is an insect moulting hormone. It is structurally different from mammalian steroids and is secreted from the prothoracic glands. Ecdysone and its homologues are generally called as **ecdysteroids.** Ecdysteroids act as moulting hormones of arthropods but also occur in other invertebrates where they can play a different role. They also occur in many plants known as phytoecdysteroids mostly as protection agents (toxins or antifeedants) against herbivore insects. A pesticide sold with the name MIMIC has ecdysteroid activity, although its chemical structure has little resemblance to the ecdysteroids.

Ecdysone agonists An agonist is a molecule that mimics the action of an endogenous biochemical molecule such as hormones and selectively binds to the same receptor and triggers a similar response in the cell, e.g. in arthropods, the generally accepted ligand for ecdysteroid receptors is 20E (20-hydroxy ecdysone), but other ligands are also recognized by these receptors. Two classes of ecdysone agonists available in plants are as follows:

Ecdysteroid compounds Plants are a rich source of agonists of 20E, e.g. muristerone*A*, isolated from *Ipomeoca calonyction*. Crop plants appear to not contain significant levels of these compounds.

Non–steroidal compounds Steroidal compounds are complex, expensive to produce and hydrophilic in nature, and detoxification processes in insects are well adapted to deal with them. Despite efforts to discover non-steroidal chemistry mimicking the activity of ecdysone, little advance has been made until recently. Since ecdysteroid receptors play an important role in the regulation of arthropod development, they are seen as appropriate targets for new pest-control agents. Bisacylhydrazines are non-steroidal ecdysteroid agonists (Wing, 1988) that are being commercialized as insecticides and other analogues are being considered as gene-switching elicitors. RH 5992 is a highly substituted member of dibenzyl hydrazine with high activity in lepidopteran species. A number of compounds from the dibenzyl hydrazine chemistry have been shown to have insecticidal activity. These compounds bind with high affinity to ecdysone receptors from the insects. These compounds when applied to growing larvae cause premature head encapsulation, preventing feeding which leads to death, e.g. RH 5992, RH 0345, RH 2485, DTBHIB, etc. They differ with different spectrum of activities. The use of dibenzyl hydrazines to control gene expression offers advantages over ecdysteroidal compounds as they are non-phytotoxic yet stable enough for use in the field.

Construction of Ecdysone Receptor Switch (ERS)

The ecdysone receptor (EcR) presents an approach to design a novel inducible system for plants. The system described the use of dexamethasone, a steroid compound, which is unsuitable for field use. The system is based on two components (Figure 9.8).

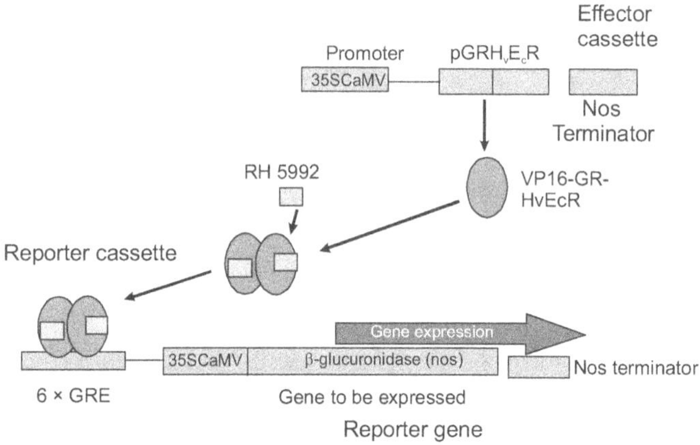

Figure 9.8 Transcriptional control of reporter gene using a chimeric nuclear receptor

1. *Effector cassette* This consists of a sequence coding for the transactivation (VP16) and DNA-binding domain of the glucocorticoid receptor combined with the sequence of the ligand-binding domain of the ecdysteroid receptor (EcR) from *Heliothis*.

2. *Reporter cassette* This consists of the reporter gene *gus* β-glucuronidase placed under the CaMV promoter and the glucocorticoid response elements (GRE) placed upstream of it.

On expression of the effector cassette, a chimeric receptor containing transactivation and DNA-binding domain of GR and ligand-binding domain of *Heliothis* EcR was produced. These chimeric proteins can bind with the ecdysone agonist, RH 5992, and subsequently with GRE of reporter cassette and can give rise to an effective expression of GUS. The efficiency of expression is due to the 6 copies of GRE at the *cis*-construct along with 35S CaMV promoter.

Applications of EcR System

To conclude, this two-component system consists of an effector which comprises a chimeric receptor containing the glucocorticoid receptor transactivation and DNA-binding domains fused to the *Heliothis* ecdysteroid receptor ligand-binding domain. The second response component contains the gene of interest under control of a chimeric promoter with six copies of the glucocorticoid response element fused to a minimal CaMV 35S RNA promoter. In the presence of a suitable ecdysone agonist, the system is activated, and transcription of the gene of interest is initiated from the chimeric promoter. The ERS system is specific to ecdysone agonists and environmental factors like drought, waterlogging, and high and low temperature. (Note: Wounding does not trigger activation of the system). The ERS provides the basis for a plant-inducible system, which has broad utility for both basic research and commercial applications.

TISSUE–SPECIFIC, COPPER–CONTROLLABLE GENE EXPRESSION IN PLANTS

The copper-controllable system makes use of the yeast copper metallothionein regulatory system. In *Saccharomyces cerevisiae*, this consists of a constitutively expressed metallo-responsible transcription factor, targeted to the nucleus, which activates yeast metallothionein transcription (Figure 9.9). This activation is mediated by copper ions that alter the conformation of the transcription factor allowing it to bind its cognate binding site in the metallothionein promoter, thereby activating transcription.

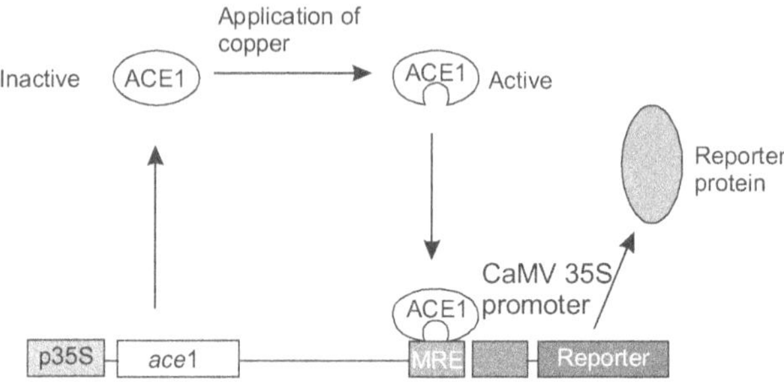

Figure 9.9   The copper-MT regulatory system

The copper-controllable gene expression system has been shown to give a tight control over time and place of expression of a "gene of interest" in response to the application of copper to transformed plants.

Copper–Metallothionein Regulatory System

The copper-controllable gene expression mechanism is based on the yeast copper metallothionein (MT) regulatory system with copper ion as the inducer molecule. This system consists of a constitutively expressed transcription factor called the metallo-responsive factor (ACE1), coded by the copper metallothionein expression-1(*ace*1) gene. When this factor is targeted to the nucleus, it activates yeast MT gene transcription (Wright *et al.*). This activation is mediated by copper ions. The conformation of the regulatory protein (ACE1) is altered allowing it to bind to specific upstream sequences called metal regulatory elements (MRE) in the yeast MT promoter. This binding is controlled by the copper ion concentrations.

Construction of the copper–inducible expression system   To translate the copper-metallothionein regulatory system into a plant background, a constitutive expression of *ace1* gene is required. So, a chimeric promoter is constructed with *ace1* gene to drive the expression of the "gene of interest" (tested with a reporter primarily) consisting of some basic plant-compatible TATA sequence together with the binding site for the ACE1 protein (Figure 9.10).

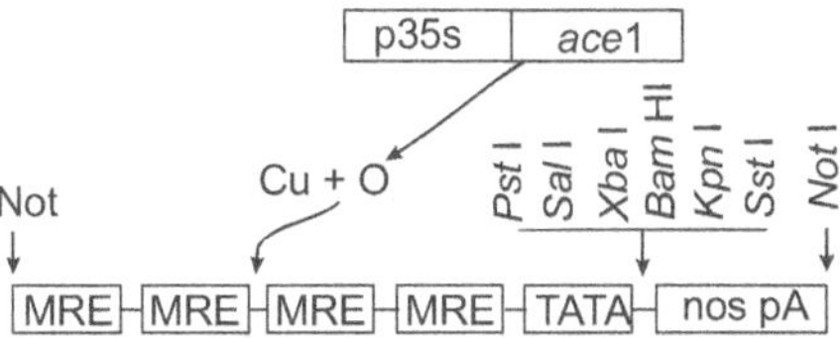

Figure 9.10 Copper-controllable gene expression system: pUC-based vector allowed cloning of the gene of interest behind the desired chimeric promoter. It has four tandem copies of the metal regulatory element (MRE) and the 46-bp fragment of the CaMV 35S RNA promoter (TATA)

Figure 9.10 shows the *ace*1 gene, placed under the control of the CaMV 35S RNA promoter (p35s). The reporter gene is placed under a chimeric promoter, which contains the MRE sequence along with sequences necessary for transcription in plants. The ACE1 protein activates transcription of the "gene of interest" (reporter) by binding to its cognate binding site (MRE—the metal regulatory element). The competence of the ACE1 protein to bind the MRE is mediated by copper (inducers), which alters its conformation to allow effective binding and activation of transcription from the chimeric promoter.

The plants with control-constructs showed a GUS-activity of 40 units/mg. In contrast, the full-construct plants, grown in the presence of 50 μmol $CuSO_4$ for 5 days, showed an increase in leaf GUS-specific activity to 1200 units/mg. GUS induction was also achieved by a foliar spray of 50 μmol $CuSO_4$. Maintenance of plants for extended periods in the presence of the "inducing" copper ion concentration in the nutrient solution resulted in the development of copper toxicity symptoms. This problem could be circumvented by the application of copper ions in a foliar spray.

The concentration of copper required in these sprays was considerably lower at 0.5 µmol.

Use of the Copper System to Effect Control Over Place and Time of Expression

Rice *et al.* (1993) used the promoter of the *nod*45 gene from lupin for tissue-specific copper-controllable expression in the nitrogen-fixing nodules of leguminous plants. The GUS-expression cassette contained a chimeric promoter with four copies of the MREs. This was cloned into a plasmid containing the *ace*1 gene under the control of the nod45 promoter. Here, the activity of the promoter was shown to be associated strictly with developed nodules of the copper-treated plants. The copper-inducible expression system showed a very low background activity in the uninduced state in the leaves of transgenic plants.

Use of the System for Tissue–specific Antisense Experiments

Functional use of the system was demonstrated by its ability to derive effective antisense constructs in an organ-specific manner. The nodule-specific system was used to express antisense constructs of aspartate aminotransferase-P2 (*AAT-P2*) in transgenic *Lotus corniculatus* plants (Mett *et al.*, 1996). Aspartate aminotransferase plays a key role in plant carbon and nitrogen metabolism. This assimilates the ammonia produced by the nitrogen-fixing process. Organ-specific expression of the antisense construct of this enzyme results in the prevention of ammonia being utilized by either bacteroids or roots, aiding in the translocation of ammonia to aerial parts.

Antisense constructs derived from the lupin *AAT-P2* cDNA were expressed in transgenic *L. corniculatus* plants. A significant antisense effect on AAT-P2 enzyme activity was seen in one of the transgenic plants with only 18% of the total AAT-P2 activity. This was observed after exposure of the transgenic plants to copper. Two other plants had only 7 and 9% of the enzyme activity upon copper induction. Thus the transgenic plants showed large decrease in nodule asparagine concentration. Prevention of NH_4 assimilation and decreased asparagine formation in root nodules helps the transport of NH_4 into the shoot where it is utilized in the formation of glutamate, one of the major amino acids required by plants. Thus overall growth of transgenic plants sprayed with copper was better compared to the wild type and transgenic plants without copper-spray.

Use of the System to Control Expression of Plant Hormones

The cytokinin group of plant hormones regulates aspects of plant growth and development including the release of lateral buds from apical dominance and delay of senescence.

Uncontrolled cytokinin expression results in an aberrant morphology of transgenic plants, whereas control of expression using the copper system allowed recovery of morphologically normal plants under non-inducing conditions.

Tobacco plants transformed with a construct containing the isopentenyl transferase (*ipt*) gene under the control of the copper-inducible promoter (Cu-ipt) was morphologically identical to the wild-type plants, under non-inducible conditions.

Delayed-senescence phenotype was observed in seedlings transformed with a copper-controlled *ipt* construct, grown in the presence of copper. Following 97 days of treatment with 5, 10 or 50 μM of copper, the leaves of the Cu-ipt plants were greener than the control Cu-GUS plants, under the same copper treatment. Only traces of cytokinin were present in GUS-positive plants (control) whereas the copper-treated Cu-ipt plants showed a total of 132 pmol/g (fresh weight) of cytokinin concentration.

To conclude, this mechanism has been translated into a plant background in a system in which there is constitutive expression of the *ace*1 gene and control of expression of the 'gene of interest' from a chimeric promoter consisting of a minimal promoter fused to the cognate binding site of the ACE1 transcription factor. In the presence of copper ions, the ACE1 protein is able to bind the chimeric promoter and activate expression. Tissue specificity was introduced into the copper-controllable gene expression system by the use of a tissue-specific promoter to effect spatial control of the expression of the ACE1 transcription factor (Mett *et al.*, 1996).

CLASS II PROMOTERS

NITRATE INDUCIBILITY OF GENE EXPRESSION USING THE NITRATE REDUCTASE GENE PROMOTER

Nitrogen is the main constituent of important macromolecules such as proteins, nucleic acids and chlorophyll. It is one of the essential nutrients required for plant growth. Plants utilize nitrogen in the form of nitrate. The sources of this inorganic nitrogen are atmospheric nitrogen and ammonium. Ammonium is rapidly nitrified to nitrate by soil bacteria.

Plant roots absorb nitrate in the environment by the action of specific permeases. Nitrate is reduced to nitrite by nitrate reductase (NR) and nitrite is converted to ammonium by nitrite reductase (NiR) (Figure 9.11). Incorporation of ammonium into organic compounds occurs by the action of the enzymes, glutamine synthetase (GS) and glutamine 2-oxoglutarate aminotransferase (GOGAT). The end products of assimilation are glutamine and asparagine.

Nitrate assimilatory pathway is induced by nitrate. Nitrate induces expression of GS and GOGAT. Light is required for optimum induction of NR and NiR gene expression in the presence of nitrate in photosynthetic tissue.

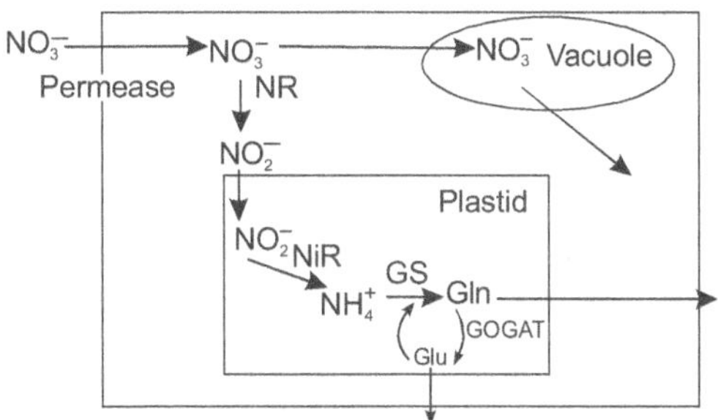

Figure 9.11 Schematic representation of the nitrate assimilatory pathway in plants

The products of *nit2* and *nit4* play a regulatory role in nitrate metabolism. The structural genes for NR and NiR are *niaD* and *niiA* respectively in *Aspergillus nidulans* and *nit3* and *nit6* code for NR and NiR respectively in *Neurospora crassa*, NIT2 and NIT4 DNA-binding domains that can turn on the expression of several nitrogen structural genes. In *N. crassa*, the *nit4* gene controls the nitrate-induced expression of nitrate reductase (NR) and nitrite reductase (NiR). The promoter of *nit3* gene is 1.3 kb in size and has three NIT2-binding sites and two NIT4-binding sites. *nit3* cannot be transcribed singly by either NIT2 or NIT4 but only by the combined presence of the two. Hence it is presumed that there may be a protein–protein interaction between NIT2 and NIT4.

Another nitrogen regulatory gene, *nmr*, in *N. crassa* appears to negatively act by preventing the expression of NR, NiR and other N-related genes in the presence of sufficient ammonia or glutamine (Premkumar *et al.*, 1980).

In nmr mutants, the nitrogen structural genes are constitutively expressed. The *nmr* gene itself is not subject to regulation and is constitutively expressed.

Nitrate Inducibility of Gene Expression in Higher Plants

In higher plants, the nitrate uptake system, NR and NiR are induced as a primary response to environmental nitrate. In addition to these, nitrate also induces the expression of GS and GOGAT. Environmental nitrate leads to the following series of events:

i. transport of nitrate from root to shoot

ii. proliferation of root tissue

iii. changes in root to shoot growth ratios

iv. enhancement of respiration.

Upon addition of nitrate to plants, there is a short lag phase, followed by a rapid increase in NR mRNA. Following the reduction of nitrate to nitrite by NR, NiR catalyses the six-electron reduction of nitrate to ammonium, using reduced ferredoxin as reductant in the chloroplasts of green leaves (Wray, 1993).

In spinach, nitrate and light regulate the level of NiR transcripts (Neininger *et al.,* 1993). A series of five deletions was made in the 3.1-kb upstream region of the NiR gene and each of them was fused to the GUS reporter gene and expressed in transgenic tobacco. GUS activity was induced in the presence of nitrate in all deletion constructs between −3100 and −330 (Rastogi *et al.,* 1993) (Figure 9.12).

These deletions did not affect the nitrate-induced tissue-specific expression of GUS, which in all cases is confined to mesophyll cells in leaves and vascular tissue in stem and roots.

Induction of GUS activity in roots occurred within 2 hours after first exposure to nitrate. In shoot, GUS activity was observed within 6 hours of exposure to nitrate.

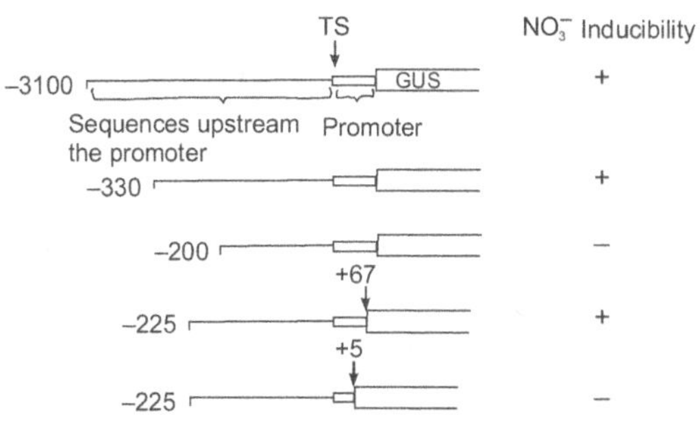

Figure 9.12 Schematic representation of 5′ and 3′ deletions of the spinach NiR promoter and nitrate inducibility of the GUS gene under the control of these promoter deletions. TS indicates transcription start site, + indicates induction by nitrate and − indicates lack of induction by nitrate.

When the NiR gene promoter is deleted to –200, the nitrate-induced expression of GUS is lost indicating the importance of the elements between –330 and –200 that mediate the induction of gene expression in response to nitrate.

Further deletion analysis of the NiR gene promoter between –330 and –200 revealed two adjacent GATA elements separated by 24 base pairs located at –214 to –211 and at –186 to –183 (Figure 9.13). NIT2 protein binds to these GATA elements of NiR gene promoter. NIT2 also binds to two GATA elements of NR gene promoter. Interaction of NIT2 with the promoters of NR and NiR genes mediate nitrate-induced gene expression. In *Lycopersicon esculentum*, nitrate-induced β-galactosidase activity was observed when the NIT2-βGAL construct was used (Jarai *et al.*, 1992).

```
-294  TCAGTCCTTTAACATTAAATCAAAACCTAGTTAGTTTTTCATACATACATTGATCAATTT

-234  TATGCAAGCGACAAAAATAGATATTAGTAACACAACATTACATATTATATCTAGCACCA

-174  TACTATAATGGTTGGCGGCTAGAGGCAGTCTGCCCTTTTAGCCGCTAGTTTTGGGTGTGA

-114  ATGGCCATCCAACGTAACCAAACATACAAAATGACCCTTAACCATGTCCAAGAGTCCCCT
```

Figure 9.13   The 5′upstream region of the spinach NiR gene promoter between −54 and −294, which harbours the two GATA elements (boxed) and the A(C/G)TCA sequence motifs (underlined)

Construction of a molecular switch using the promoter region of nitrate reductase is still in its infancy.

HEAT-SHOCK PROMOTERS TO CONTROL GENE EXPRESSION

Heat-shock proteins (HSPs) are proteins that are induced at different temperatures in different organisms. In plants, the HS response occurs after an elevation of approximately 8–12°C above the normal growing temperature and is characterized by a very rapid induction of HS gene transcription with a concomitant decline in the transcription of most other genes. High temperature stress is one of the many different stresses that plants and other organisms encounter. HS response was first observed in *Drosophila* and occurs in almost all organisms ranging from bacteria and lower eukaryotes to mammals and plants.

In agricultural settings, the system inducer, heat, is naturally present without additional input or adverse environmental impact often associated with the application of chemicals. Continued expression is assured by repeated activation of the system on a daily basis by the natural diurnal temperature cycle of the growing season. HS promoter system offers an alternative to the present system that is reliable, economical and virtually maintenance-free.

Most organisms including plants can survive at lethal temperature treatment if they are pretreated with a non-lethal high temperature, leading to HSP synthesis. This phenomenon is called **induced or acquired thermotolerance**. The HS response is the transient nature of HS gene expression, ranging from a few minutes in *E. coli* to a few hours in higher plants.

Based on their molecular weights, the HSPs are designated as HSP100, HSP90, HSP70, HSP60 and low-molecular weight (LMW) HSPs (15–30 kDa). The LMW HSPs have been classified into at least 6 gene families targeted to different cellular components including the cytosol, chloroplasts, mitochondria and endoplasmic reticulum.

Organization and Types of Heat–Shock Promoters

There are basically 3 types of heat-shock promoters based on their organization. Class A promoters are primarily regulated by the binding of activated HSF under heat-stress conditions. Auxiliary elements such as AT-rich sequences enhance the amplitude of heat induction. Class B promoters are also dependent on HSEs for activity, but require either specialized HSFs that are active under non-HS conditions, analogous to HSF2 in mammals, or tissue-specific recruitment of the normally stress-responsive HSF. Class C promoters also usually contain HSEs but have other types of elements that independently activate transcription of HS genes under non-stress conditions.

Heat induction of gene expression is dependent on the presence of HS consensus elements (HSEs) in the promoter (Pelham, 1981) (Figure 9.14). A trimer of the HS transcription factor (HSF) binds to these elements after the cell has experienced a heat stress, and strongly activates the transcription of HS genes. There is a consensus **HSE core** (5′-GAA-3′) in soybean Gmhsp 15.5 (Gm-Glycinemax) promoter, which is preceded by an A and followed by G to give the sequence 5′-aGAAg-3′ or 5′-cTTCt-3′ depending on the orientation. For optimal function, an HSE must contain at least four recognizable HSE cores located 2 bp apart in an alternating array (5′-GAA nn TTC nn GAA nn TTC-3′). The HS inducible promoter requires the binding of at least two of the subunits of HSF trimer for efficient induction. There is usually at least one perfect HSE core within the 30-bp upstream of TATAA embedded within a cluster of imperfect HSE cores to comprise site I (Class A). For full activation, HSFs must occupy both site I and site II cluster located within approximately 120 bp from the transcription start site (Gurlev and Key, 1991).

The HS promoter of class A type is regulated by the binding of four trimers of the HSF to HSE sites I and II (Figure 9.15). Cooperative interactions between adjacent HSF trimers facilitate the binding of these trimers with imperfect HSE cores of sites I and II.

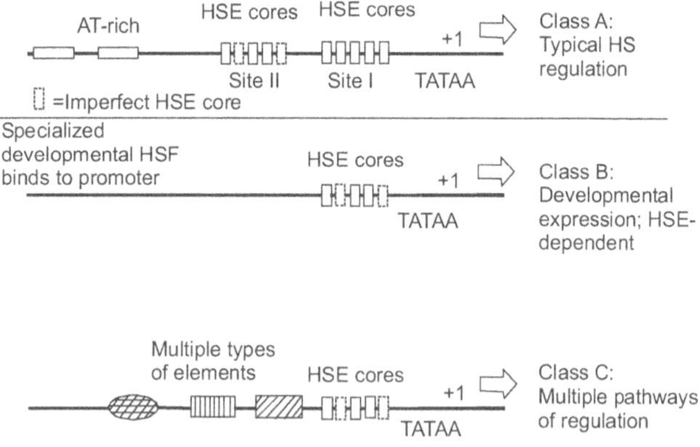

Figure 9.14 General classes of heat-shock (HS) promoters

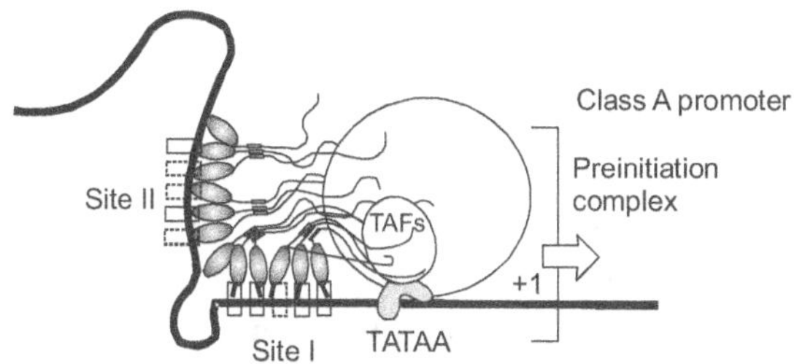

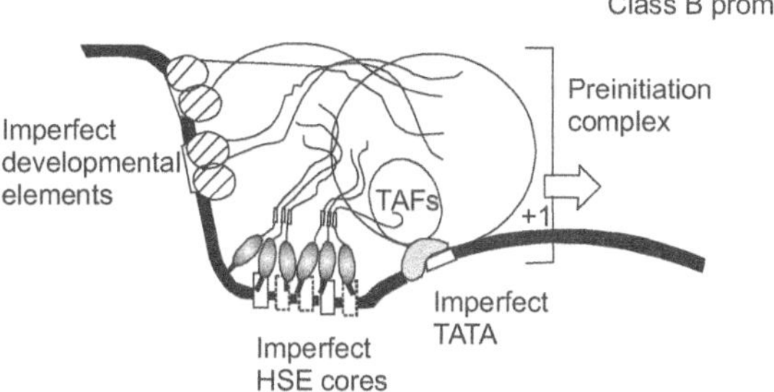

Figure 9.15 Hypothetical interactions of activator proteins and general transcription factors with two types of heat-shock promoters.

The Class B promoter is dependent on cooperativity in binding, involving three components.

1. HSF trimers
2. Developmental factors
3. The preinitiation complex

The affinity of each component alone for its respective binding site on the DNA is insufficient for promoter activation. Each component stabilizes the DNA-binding of the other two components.

Class A promoters are primarily regulated by the binding of activated HSF under **heat-stress conditions**. Auxiliary elements such as AT-rich sequences enhance heat induction.

Class B promoters are also dependent on HSEs for activity, but require either specialized HSFs that are active under **non-HS conditions** and under stress.

Class C promoters also usually contain HSEs and other types of elements that independently activate transcription of HS genes under **non-stress conditions.**

In plants, the distinction between classes of HS promoters is not always clear. Many plant HS promoters can be tentatively placed in class A category, since they appear to be induced only by heat and related stresses, and seem to contain no recognizable consensus element in the promoter, e.g. promoters for the LMW HSPs of soybean and other plants. These genes are also induced in vegetative tissues by a variety of other stresses.

The sunflower Ha Hsp 17.6 G1 gene is expressed during seed development but is non-inducible by heat stress (Carranco *et al.*, 1997). Although the Ha Hsp 17.6 G1 promoter is not inducible by heat-stress, a cluster of perfect and imperfect HSE cores is located approximately 50-bp upstream the putative TATA box, a position, which is comparable with site II in Class A promoter.

There are four HSFs in mammals. Of the four, only HSF1 is primarily specialized for responding to heat-stress. HSF2 and HSF3 are involved in the developmental expression of *hsp* genes. HSF4 seems to be a regulatory transcription factor that shuts down the HS genes under non-stress conditions (Nakai *et al.*, 1997). These positive and negative regulatory elements are considered in the design and use of HS promoters to engineer control of gene expression.

Few studies have quantitatively compared the strength of different constitutive and inducible promoters. Holtorf *et al.*, (1995) compared CaMV 35S, CaMV 35–omega, *Arabidopsis* ubiquitin UBQ1, TMV and soybean HS promoter in transgenic plants using the GUS reporter gene. The CaMV 35 promoter had the highest expression followed by UBQ1, TMV promoter and finally soybean HS promoter. In *Arabidopsis*, the HS promoter fused to the GUS reporter gene showed constitutive floral-organ-specific expression under normal growth temperature (22°C).

Role of HSP in Plants

Many experiments have shown that the acquisition of thermotolerance correlates with conditions resulting in synthesis of HSPs. For example, a deletion mutant lacking the *HSP*104 gene failed to acquire thermotolerance; however, reintroduction of the *HSP*104 gene rescued the thermotolerant phenotype. These results demonstrate that at least one HSP plays a critical role in cell survival at extreme temperatures. Although HSPs were initially defined as proteins whose expression is highly induced by elevated temperature, many recent studies indicate that these proteins are also regulated by a variety of environmental and developmental signals in animals. This supports two ideas regarding the role of HSPs.

   i.   HSPs can be functionally distinct between classes and even between members of the same class and

  ii.   HSPs so induced provide functions essential to the developmental process. The promoters regulating these genes (HS promoters) must therefore be functionally distinct, either containing developmentally activated motifs or unique configurations of HS elements.

HSP in Pollen Development

In general, mature pollen of most plant species lacks a normal HS response. In several species, no HSPs (heat-shock proteins) are synthesized in response to HS. An exception is the pollen of

Sorghum where immature pollen and *in vitro* germinating pollen synthesize HSPs at heat-stress temperatures (Frova *et al.*, 1991). In maize, mRNA transcripts encoding 18-kDa HSPs are expressed in a stage-specific manner during microsporogenesis without heat stress. The genes encoding two 18-kDa HSPs are expressed and accumulated independently during microsporogenesis implying gene-specific developmental regulation rather than general activation of the HS or stress response. In tobacco pollen embryogenesis, Northern blot analysis showed that expression of a class I LMW-HSP, Nt Hsp, is activated at normal temperatures during the dehydration phase of pollen development just before anthesis. This implies that the expression of LMW-HSPs plays an important role in pollen development.

There is a lack of HSP induction in pollen for many temperate crop species. HS treatment of germinating maize pollen does not induce HSPs.

HSP in Embryo Development and Seeds

HSP70-genes have been shown to be differentially expressed during development. In pea leaves, PsHSP71.2 is strictly heat-inducible. PsHSP70b is constitutively expressed at low levels, but strongly induced by heat stress. There is a selective activation of HSP gene expression and a precise activation of specific HSP promoters under non-heat stress conditions.

Zimmerman *et al.* (1989) reported the presence of abundant HS mRNAs in the calli of carrots before heat-stress. Upon heat-stress, callus cells translate only a fraction of the abundant HS mRNAs they accumulate.

Construction of Heat–Inducible Expression Cassette

A heat-inducible expression cassette was constructed using the soybean *Glycine max* HSP promoter, to study the conditional expression of any sequence of interest in transgenic plants or plant tissues (Ainley and Key, 1990). The heat-inducible production of a cytokinin in transgenic tobacco plants was tested by insertion of an isopentenyl transferase (*ipt*) gene into this HS cassette. Heat-induced synthesis of endogenous cytokinin produced several beneficial effects.

Heat treatment of 1–2 h/day for 1–4 days was sufficient for observable phenotypic changes in 3–4 weeks after the heat treatment, indicating that the transient heat induction is sufficient for expression.

Heat-inducible hygromycin was attained in transgenic tobacco by a construct consisting of the soybean GmHsp17.5 L promoter fused to a hygromycin phosphotransferase gene (Severin and Schoff, 1990). Incubation for 1 h at 40°C everyday, applied for several weeks, was sufficient to express a hygromycin-resistant phenotype.

One potentially very important use of HS gene promoters would be the expression of genes that are normally shut down during periods of high temperature stress. It has been reported that synthesis of PR-proteins is suppressed upon shifting to elevated temperature. Pathogen resistance of many crop plants may be improved by transformation with HS promoter/PR protein fusions to induce expression of PR-proteins which are normally repressed by the HS-response.

To conclude, LMW HSP promoters would be suitable to achieve the highest inducible expression with minimal constitutive expression. If developmental expression is undesirable, a strictly heat-inducible HSP70 promoter may be preferred because it may have lower developmental expression than some LMW HSP promoters. For some genes, it may be desirable to have expression at normal temperatures and enhanced expression with elevated temperature. For such criteria, an HSP90 class promoter or heat-inducible ubiquitin, or the soybean Gm Hsp26 A promoter may be used.

WOUND-INDUCIBLE GENE EXPRESSION IN PLANTS

All living organisms are involved in a constant struggle with and against other organisms for survival. Plants have developed potent biochemical responses to protect their integrity and to restrict the entry of invasive pathogens. Structurally, plants have a polyester coating composed of cutin and suberin. However, if a break or wound occurs in this surface coating, then pathogens gain entry into the plant tissues and cause extensive damage. The response of plants to wounding has been studied in detail since Green and Ryan (1972) reported the accumulation of an inhibitor of chymotrypsin in tomato leaves in response to wounding. Since that time, at least 70 other proteins have been identified to be wound-inducible.

Phases of a Wound Response

Wounding results in the activation of many genes in a plant. Various steps involved in wound responses are as follows:

1. placing mechanical barriers to invade organisms
2. sealing the wounded tissue
3. activating defence compounds against invading organisms
4. recovering from the wound

The sum of these processes results in normal physiology of plants.

The initial phase of a wound-response is a rapid reaction to close the wound, thereby protecting the plant from loss of cellular components and restricting the entry of microorganisms into the plant tissues. This takes place in two steps.

1. Initially there is an immediate oxidative burst that results in a cross-linking of plant cell wall proteins (Bradley *et al.*, 1992). This oxidative burst can be detected within 15 seconds.

2. The cross-linking of the cell wall proteins provides a structural barrier that inhibits the invasion of microorganisms. In addition, hydrogen peroxide (H_2O_2) from this oxidative burst is thought to activate some of the wound-inducible genes. Since hydrogen peroxide is toxic to tissues, plants have a physiological control over peroxide accumulation (Mohan *et al.*, 1993).

The accumulation of H_2O_2 ends up in the up-regulation of genes involved in phenyl propanoid pathway such as those of phenylalanine ammonia lyase (PAL), chalcone synthase and chalcone isomerase. The induction of these genes is followed by formation of lignin precursors, which can reseal the wounded surface and provide cells with precursors of phenolic plant-defensive compounds.

The second phase of the wound response, particularly in monocots, is the turn-off of photosynthetic protein translation by arresting the transcription of nuclear-encoded photosynthetic genes. Because the maintenance of the photosynthetic apparatus represents a major expenditure of cellular energy, repressing the synthesis of new proteins would save energy for the plant following the wound. Among the down-regulated proteins are those for the small subunit of ribulose 1,5-bisphosphate carboxylase/oxygenase (RuBisCO) and several light-harvesting complex apoproteins. Several mechanisms are responsible for this altered regulation of the photosynthetic machinery. First, methyl jasmonate induces a shift in the 5′ untranslated region of the *rbcL* transcript (Reinbothe *et al.*, 1993). The primary transcript is initiated at −316 from the translation start codon. Under normal conditions, the 5′-end of the mature *rbcL* transcript is processed to yield an mRNA with a 59-bp 5′ untranslated region (UTR). Following the jasmonate treatment (which mimics wounding), the mRNA is processed to give a 94-bp untranslated region. This alternatively spliced transcript contains within the 5′UTR the 35 base motif that has high complementarity to the 3′-terminus of the 16S rRNA. This portion of the 16S rRNA is involved in intramolecular base pairings within the ribosome and can associate with 30S but not with 70S complexes, leading to down-regulation of the large subunit which in turn leads to regulation of small subunit.

A second method that plants use to alter protein synthesis in stressed plant tissues involves the expression of ribosome-inactivating proteins.

Induction of Defensive Compounds

Plants have developed a variety of biochemical defences to combat invading pathogens and small herbivores that wound them. Many different plant species have been shown to activate the synthesis of phytoalexins after a wound or after methyl jasmonate treatment. Phytoalexins are plant-synthesized small-molecular weight defensive compounds that have inhibitory action against microorganisms or herbivores. These include phenolic, terpenoid alkaloid compounds, furanocoumarin, taxol, momilactone, etc. Ramputh and Brown (1996) reported the accumulation of the inhibitory neurotransmitter GABA (gamma-aminobutyric acid), following mechanical damage of soybean leaves. The increasing levels of GABA decreased the survival of the feeding larvae and increased the length of time that the larvae required to pupate.

Leaf damage by herbivores in *Nicotiana sylvestris* produces a damage signal that dramatically increases the de novo nicotine synthesis in the roots. The increased nicotine pool is then transported up the plant. High levels of nicotine make the leaves unpalatable to herbivores. Other proteins induced by wound are serine-proteinase inhibitors (Ryan, 1981), α-amylase inhibitors (Ishimoto and Chrispeels, 1996), chitinases (Broglie *et al.*, 1991), β-glucanases (Mauch *et al.*, 1988), osmotin (Grosset *et al.*, 1990) and others. Each of these enzymes or inhibitors performs a specific function in combating the invading herbivore or pathogen. The serine-proteinase inhibitors are active against insects. These degrade the essential proteases of the insect gut. Consequently, in the absence of protease, free amino acids are not liberated and hence the insect dies.

While plants have many serine-proteinase inhibitors, the presence of serine-proteinases in plants is rare (Ryan, 1981). Thus, plants apparently lack the specific target or substrate of these inhibitors. Serine-proteinases are however present in abundance in insect guts which form the target of the plant serine proteinase inhibitors. Chitinases and β-1,3-glucanase are other defensive enzymes that have no natural target in plants. Chitin does not exist in plants and β-1,3-glucans are not major components of plant cells. Chitin and β-1,3-glucans are however extensively found in the cell walls of fungi.

Mechanism of Wound Induction

It is certain that numerous mechanisms are responsible for wound-inducible gene expression in plants because of the induction of wide number of genes that are activated in response to wounding. Among them the best characterized are the proteinase inhibitor genes of solanaceous plants and the vegetative storage protein genes that are similarly regulated.

Systemic signal The local wounding triggers the expression of proteinase inhibitor genes at a distal site in solanaceous plants. Two mechanisms have been proposed to trigger the wound-induced systemic accumulation of proteinase inhibitors. These mechanisms are mediated by either electrical or chemical signals.

Electrical signals   Wildon *et al.* (1992) showed that wounding of the cotyledons of a young tomato plant results in a slow-moving action potential that propagates away from the site of the wound towards the upper leaves. In all cases, this action potential correlates with the induction of proteinase inhibitor genes. Plants are unique, in that they have symplastic connections that continue throughout the organism. These connections are made by plasmodesmata and are well suited for electrical signals. Rhodes *et al.* (1996) showed that the electrical signals travelled from the wounded cotyledon to distant unwounded leaves along sieve-tubes and companion cells of the phloem. However, the mechanisms that translate this action potential into a chemical form that activates gene transcription have not been fully elucidated. There have been several ion channels identified in plants (Lurin *et al.*, 1996) that could possibly participate in this process. Acetylsalicylic acid is one of the inhibitors of wound-inducible gene expression known to disrupt H^+/K^+ transporters at the plasma membrane (Glass and Dunlop, 1974).

Systemin   One of the most interesting findings in the area of plant biochemistry is that the polypeptide signals may function in the activation of plant defence genes. A polypeptide from tomato leaves at very low concentrations was capable of initiating the signal transduction cascade leading to the expression of proteinase inhibitor genes in the absence of a wound (Pearce *et al.*, 1991). A synthetic polypeptide identical to the one purified from plants was also active in inducing the proteinase inhibitor gene. This synthetic polypeptide was readily mobile in the phloem, as opposed to oligosaccharide signals (Baydoun and Fry, 1985). The gene encoding the signalling molecule, systemin, has been isolated and characterized. It is synthesized from a 200-amino acid pro-protein termed prosystemin that is encoded in 11 exons. Systemin must be proteolytically processed to release the active systemin peptide. Plants transformed with an antisense copy of prosystemin cDNA showed a dramatic suppression of proteinase inhibitor expression in the leaves of the transgenic plants (McGurl *et al.*, 1992). An over-expression of prosystemin cDNA in tomato plants resulted in a constitutive expression of proteinase inhibitor proteins in leaves. Systemin is also capable of inducing other plant defensive proteins (such as polyphenol oxidase) other than proteinase inhibitor. Constabel *et al.* (1995) grafted non-transformed, wild type of scions on to the transgenic root stock and demonstrated that a signal could be transmitted from root stock transformed with the prosystemin cDNA through a graft junction to non-transformed leaves in the absence of wounding. Narvaez-Vasquez *et al.* (1994), used *p*-chloromercuric benzenesulphonic acid (PCMBS), an inhibitor of active apoplastic phloem loading. PCMBS was shown to be a powerful inhibitor of wound-induced and systemin-induced activation of proteinase inhibitor synthesis in tomato leaves.

Localized interactions Once the long-distance systemic signal reaches its local site of action, that signal (whether electrical or chemical) must be transduced to the nucleus of the cell where gene transcription occurs. Electrical signals are known to operate through ion channels in cells that could lead to the transducing chemical signal. But the involvement of such ion channels have not been demonstrated with any of the chemical signals known to induce wound-inducible genes. Thus, chemical signals interact with a cell-surface receptor that then transmits chemical energy across the membrane to the cytoplasm.

The interaction of oligosaccharide elicitors with cells leads to several alterations in the plasma membrane. It is known that wounded plant cells have increased membrane fragility perhaps due to phospholipase action. Further, elicitor treatment of cells leads to the phosphorylation of various plant plasma membrane proteins in tomato. Elicitation with systemin resulted in the hyperphosphorylation of a 27-kDa protein. This indicates that protein kinases play an important role in the mechanism of signal transduction leading to the expression of defensive genes.

Oxylipins   Jasmonic acid and its methyl ester, methyl jasmonate, are active in inducing the accumulation of numerous wound-induced proteins in plants. Leaf damage in *Nicotiana sylvestris* rapidly causes the level of shoot jasmonate pools to rise rapidly (< 0.5 hour). Root jasmonate pools also rise in response to leaf damage, but more slowly (< 2 hours). The levels of jasmonic acid remain elevated for 24 hours in shoots and 10 hours in roots (Baldwin *et al.*, 1994). The following flowchart shows the sequence of synthesis of jasmonic acid.

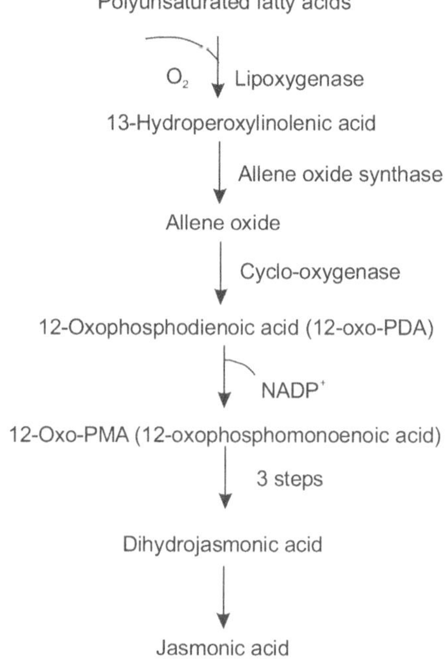

For the synthesis of jasmonic acid, the starting products (free fatty acids) have to be liberated from membrane phospholipids. It is currently unclear whether these free fatty acids are liberated from chloroplast phospholipids or from other membranes.

The fatty acids are transported to the chloroplast via lipid transfer proteins. The chloroplast envelope membranes are believed to be intracellular locations of jasmonate biosynthesis. The conversion of free fatty acids into jasmonic acid occurs in 5 steps through an oxidative pathway. The intermediates are termed as oxylipins.

Lipoxygenases in plants are wound-inducible. Transgenic plants harbouring the gene encoding allene oxide synthase accumulated high levels of jasmonate which was comparable with the high jasmonate concentrations in wounded nontransgenic plants.

Usually, jasmonate levels decline rapidly following the burst of synthesis. Krumm *et al.* (1995) observed that jasmonates conjugate with amino acids and become inactive. These authors speculate that these conjugates may function in the long-term maintenance of jasmonate in plants.

Functions of jasmonate   After jasmonates are synthesized, it is unclear how the biological activity of these compounds are transmitted to the promoters of the various genes that they activate. There is a report of a jasmonate-binding protein that mediates the wound-inducible regulation of transcription of the potato proteinase inhibitor 2 gene (*pin2* gene) (Gurevich *et al.*, 1996). Another factor that triggers proteinase inhibitor expression is present downstream of jasmonate and was discovered by Schaller *et al.* (1995). They also found another compound, 'bestatin' that induced proteinase inhibitor genes without affecting systemin, octadecantids or jasmonate.

Involvement of other Hormone Factors

ABA (*Abscisic acid*)   Wounding induces the biosynthesis of abscisic acid prior to transcription of wound-inducible genes. Hildmann *et al.* (1992) demonstrated that exogenous application of ABA induces the accumulation of proteinase inhibitor II mRNA that is identical to mechanical wounding. They also showed that ABA-deficient plants do not respond to wounding unless ABA is supplied exogenously. There is an increase in ABA levels in the leaves of tomato, potato and tobacco plants following a wound. Systemin leads to an increase in ABA which in turn leads to an increase in jasmonic acid which then regulates gene transcription. According to this hypothesis, all jasmonate-regulated genes should also be ABA-regulated. Lee *et al.* (1996), however, identified 4 genes, which are regulated by jasmonate but not regulated by ABA indicating that the signalling pathways for ABA and jasmonate function independently and not sequentially.

ABA leads to the activation of a lipoxygenase that generates hydroperoxides from free fatty acids within the cell. In addition to wounding, water-stress also causes accumulation of ABA which then activates a set of water-stress genes. However, it does not induce wound-inducible genes. Thus, ABA plays differential roles according to the type of stress it perceives.

Ethylene   Ethylene is synthesized following a wound and many wound-inducible genes are also responsive to ethylene. There is an accumulation of considerable amounts of ethylene following wounding. It is formed from 5-adenosylmethionine (with precursors being methionine and ATP), in 2 steps (Refer Chapter 5). The first step is catalysed by the enzyme amino cyclopropane carboxylic acid synthase (ACC synthase) and the second step is catalysed by ACC oxidase. These genes are usually expressed in fruits. However, each of the steps in this pathway is also wound-inducible. A set of defence genes induced by ethylene is called "ethylene-related genes". But the functions of the proteins encoded by these genes are unknown.

O'Donnell *et al.* (1996) demonstrated that ethylene is absolutely required for wound-induction of the proteinase inhibitor genes of tomato. They used norbornadiene, which is an inhibitor of ethylene synthesis, and showed the absence of proteinase inhibitor-gene expression that is essential for signalling plant defensive genes.

The tomato ethylene mutant termed "Never ripe" has a partial loss of ethylene sensitivity (due to mutation in the ethylene receptor protein). In this plant, the wound-induced accumulation of proteinase inhibitor transcripts is significantly delayed. Also, transgenic tomato plants expressing an antisense ACC oxidase do not accumulate proteinase inhibitor transcripts in response to wounding.

Auxin   Auxin has been demonstrated to prevent expression of wound-inducible proteinase inhibitors (Kernan and Thornburg, 1989). Inhibition of gene expression by auxin has been demonstrated for a number of other wound-inducible genes. Auxin has been shown to strongly inhibit methyl jasmonate-induced wound-inducible gene expression in soybean suspension-cultured cells and the expression of β-glucanase in response to fungal elicitor in tobacco and soybean cells. Thornburg and Li (1991) demonstrated that IAA declined by two to threefold following a wound and this decline is inversely correlated with the induction of wound-inducible gene expression.

GTP-binding proteins also lead to the synthesis of jasmonates. Cytokinins are essential for the accumulation of wound-inducible proteinase inhibitor transcripts.

Induction of Storage Proteins

In plant families, vegetative storage proteins accumulate in leaves prior to anthesis, decline during pod filling and then accumulate again after seed maturation. These proteins play a significant role in plants especially during the rejuvenation of axillary buds in spring. In addition to this developmental role, these proteins (which are induced by wounding and by jasmonates) help the plants to protect themselves from loss of metabolites during the wound response. The storage proteins are rich in nitrogen and carbon and serve as a source for new growth after the wound recovery phase.

Thus the involvement of multiple long-range signals, several signal transduction cascades involving kinases and phosphatases along with variations in multiple plant hormones, is clear. Ethylene, cytokinins, auxins and ABA and the biosynthesis of jasmonates play a role in the transcriptional activation of wound-inducible genes.

SENESCENCE–SPECIFIC PROMOTER IN DEVELOPMENTAL TARGETING OF GENE EXPRESSION

The terminal developmental phase in the life cycle of a plant is generally referred to as senescence. Senescence also refers to the fall of leaf, flower and fruit. Leaf senescence is a genetic programme influenced by many external and internal factors.

Leaves are the primary sites of photosynthesis. During senescence, anabolic processes are replaced by catabolic reactions such as degradation of chlorophyll and leaf proteins (approximately 60% of leaf proteins are degraded). Although leaf senescence recycles nutrients in soil, it reduces yield in certain crops like soybean and maize. In addition, a senescing leaf becomes more susceptible to pathogen attack.

Role of Cytokinins in Plant Senescence

While there are many processes leading to leaf senescence, there are few factors that retard senescence. Cytokinin, a phytohormone, was found to inhibit leaf senescence. Cytokinins can be added exogenously to plants to prevent senescence or the plants can be transformed to produce high levels of endogenous cytokinins. The isopentenyl transferase (*ipt*) gene of *Agrobacterium tumefaciens* that is involved in cytokinin production has been cloned. This gene is expressed in plants with a variety of promoters such as heat-shock and light-inducible promoters. With high levels of cytokinins, the transgenic plants showed delayed senescence. If a leaf senescence-specific promoter is used to direct the expression of the cytokinin-synthesizing gene, *ipt*, cytokinin production in the transgenic plant is triggered at the onset of senescence. This prevents senescence and helps in agricultural improvement.

Genes activated during senescence termed as senescence-associated genes (*SAGs*) were identified following the techniques given below:

1. Construction of cDNA libraries using mRNAs of senescent tissues

2. Making duplicate sets of filters of the cDNA library and

3. Hybridization of one set of the filters with cDNA probes made from young, non-senescent tissues and the other set with senescent cDNA probes.

Comparison of the two sets of filters allows the identification of cDNA clones that hybridize only to "senescent" cDNA probes but not to non-senescent probes and therefore reveals the mRNAs that increase during senescence. Using this technique, six cDNA clones designated *SAG*12 to *SAG*17 have been identified (Lohman *et al.*, 1994). Further studies revealed that *SAG*12 and *SAG*13 were expressed only in senescing leaves and the other *SAGs* were expressed at low levels in young leaves and at high levels in senescing leaves.

The gene organization of *SAG*12 and *SAG*13 has been characterized. *SAG*12 is a single copy gene, which consists of two exons and one intron. It codes for a protein that belongs to the super family of cysteine proteinases. This family also includes genes involved in programmed cell death (PCD). *SAG13* consists of 4 exons and 3 introns. It codes for the enzyme alcohol dehydrogenase. The promoter regions of both *SAG12* and *SAG13* were fused to β-glucuronidase (*GUS*) reporter gene and were expressed in *Arabidopsis* and tobacco plants. These promoters directed GUS expression only during senescence. The signal transduction pathway inducing *SAG12* and *SAG13* is not known.

The Role of IPT in Cytokinin Production

The reaction between isopentenyl phosphate and AMP to form isopentenyl adenosine ribotide that is then converted to various cytokinins is catalysed by the enzyme isopentyl transferase (IPT). The gene *ipt* was transferred to *Arabidopsis*, cucumber, tomato and potato plants using *Agrobacterium*. These transgenic plants that overproduced cytokinin, had stunted growth, smaller leaves and impaired root growth. The promoters used to direct the expression of *ipt* were individually inducible by heat, light and tetracycline.

Use of SAG Promoters to Target *ipt* Gene to Senescing Leaves

ipt gene was expressed under the regulation of *SAG12* promoter (Figure 9.16). A NOS terminator (the terminal sequence of the nopaline synthase gene) was used as the termination signal. The expression of this construct in plant cells controlled cytokinin levels by autoregulation. That is, at the onset of senescence, the *SAG12* promoter directs the expression of *ipt* in senescing leaf cells. The increased IPT enzyme activity results in elevated cytokinin production, which prevents the leaf from senescing.

Increased cytokinin levels suppress *SAG12* promoter. This suppression prevents cytokinin overproduction.

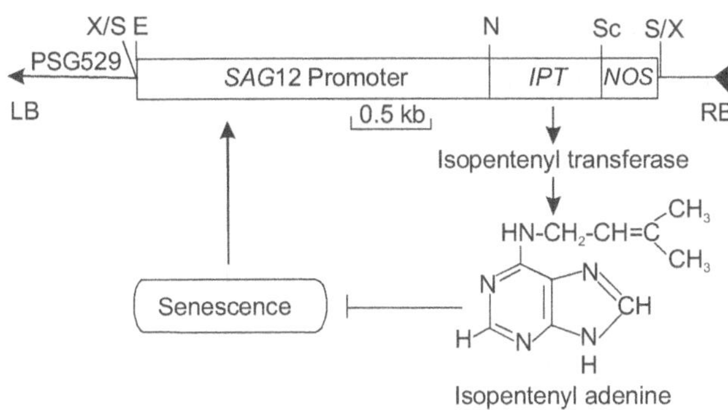

Figure 9.16 The use of *SAG*12 promoter to target IPT expression in senescing leaves

Phenotype and Physiology of *SAG*12–IPT Transgenic Tobacco Plants

The transgenics were normal except for delayed leaf and floral senescence. Seeds were viable. There were no statistical differences in overall plant height and leaf number. Although the plants grew normally, senescence was delayed in SAG12-IPT transgenic plants. This difference in senescence between SAG12-IPT transgenic and wild type plants can be observed in as early as 3.5-week-old seedlings. The cotyledons of wild type seedlings were already senescing, whereas those of transgenic plants did not show any signal of senescence of cotyledons.

At the end of 12 weeks, older leaves in wild-type tobacco plants started drooping whereas, there was no visible sign of even yellowing of petioles in SAG12-IPT plants.

At the end of 20 weeks, flowering was terminated after the production of approximately 180 flowers in wild type. In contrast, SAG12-IPT plants were still flowering and finally produced more than 320 flowers.

Not only was the flowering period prolonged in SAG-IPT plants but also the petals of individual flowers were non-senescent. In wild-type, the petals remain in flower for 3 days, whereas in transgenic plants, they persisted up to 4.5 days.

SAG12-IPT transgenic plants showed about 50% increase in both seed yield and dry weight accumulation in comparison to those of wild-type plants (Gan and Amasino, 1995).

Later, the *SAG12*-IPT plants were grafted on wild type rootstocks. The reverse was also done. That is, the wild plants were grafted on SAG12-IPT rootstocks. In both cases, normal senescence was restricted to the wild type plant region alone showing that the cytokinin produced in the transgenic leaves is not translocated upwards or downwards in sufficient amounts to cause delay in leaf senescence in wild plant regions.

Conclusion

Leaf senescence is thought to play an important role in the evolution of fitness by recycling nutrients from dying cells to actively growing parts such as reproductive organs. Senescence also devalues ornamental plants and vegetables during transportation and storage. Hence, the use of controlled *ipt* expression may result in agricultural benefits like preservation of proteins for longer periods, increase seed yield and plant biomass have been increased.

ABSCISIC ACID–AND STRESS–INDUCED PROMOTER SWITCHES IN THE CONTROL OF GENE EXPRESSION

Field-grown plants are constantly under unfavourable environmental conditions such as drought, flooding, extreme temperatures, high salt concentrations, heavy metals, and infection by pathogenic agents. The genetic programme in normal plants is altered by stress stimuli to produce specific proteins and to activate biochemical pathways, which are essential for survival. It has been established that drought, cold and salinity stresses enhance the synthesis of phytochromes and abscisic acid.

Genes Regulated by ABA

ABA is known to regulate the expression of a variety of genes, including those encoding seed storage proteins.

Four sets of ABA-regulated genes have been categorized:

1. ABA-inducible genes
2. ABA-suppressible genes
3. Genes mediating ABA responses
4. ABA biosynthetic genes.

The following are the notable enzymes whose genes are regulated by ABA:

1. aldose reductase in barley
2. glyceraldehyde 3-phosphate dehydrogenase in rice
3. sucrose-phosphate synthase
4. L-isoaspartyl protein methyltransferase in wheat
5. serine/threonine protein kinase
6. thioprotease in *Arabidopsis*
7. peroxidase in duckweed

Mechanism of ABA Action and ABA–Responsive Promoters

Mundy *et al.* (1990) have reported that a promoter fragment, between –294 and –52 of the rice Rab (response to ABA) 16A gene is sufficient to confer ABA-inducible expression of the chloramphenicol acetyltransferase (CAT) gene in rice protoplasts.

Sequence comparisons of ABA-inducible genes show the presence of the consensus ACGT-core. This type of sequence containing an ACGT core has been named the ABA response element (ABRE). This ACGT core containing ABRE is conserved in almost all the ABA-regulated genes. This ABRE core sequence is similar to the conserved G-box motif found in a number of yeast promoters and plant promoters and E-box found in the promoter of certain mammalian genes.

In barley (*Hordeum vulgare*) aleurone layers, ABA induces dozens of genes and two of them, HVA1 and HVA22, have also been shown to be induced by drought, salinity and temperature stress conditions. The promoters of these genes have been analysed by linking them to the coding region of the reporter gene, *GUS*, followed by analysis of GUS expression in barley tissues which have been transformed with these gene constructs by particle bombardment (biolistic technique). When the protoplasts were treated with ABA, high levels of HVA22 mRNA and GUS expression were observed (Figure 9.17).

The DNA construct containing the GUS gene driven-promoters was delivered into the aleurone cells of barley embryo less half-seeds by biolistic method. After treatment with or without ABA, the bombarded seeds were homogenized and GUS activities were determined. The test promoter in this construct was the Amy64, barley α-amylase gene minimal promoter. ABA does not induce this by itself when transformed into barley. The addition of a 49-bp promoter fragment from the HVA22 gene in either orientation gives a high level of induction. This 49-bp fragment contains an ACGT box. Similarly, a 68-bp promoter fragment from the HVA1 gene was also able to confer a high level of ABA induction. This region also contains the ACGT box.

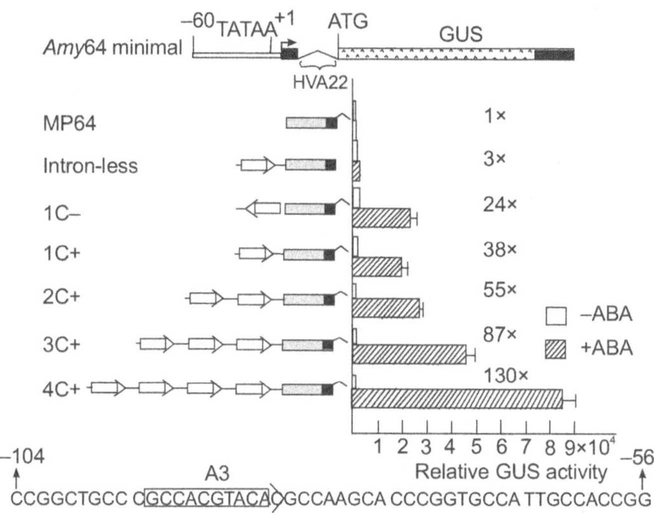

Figure 9.17 A 49-bp fragment containing A3 confers ABA inducibility to a minimal promoter. The minimal promoter (+1 to –60) and the 5′ untranslated region (+1 to +57) from the barley Amy64 a-amylase gene was fused to the 5′-end of HVA22 intron1–exon2–intron2 fragment (thin black angled line) (Shen and Ho, 1995). The 3′-region (black bar to the right of the GUS coding sequence) was from the HVA22 genomic fragment, including the polyadenylation sequence (AATAAA). This minimal promoter (MP64) is not responsive to gibberellin or ABA. The 49-bp HVA22 promoter fragment, shown at the bottom, was fused in either positive (i.e., the same orientation as in the native promoter) or negative orientation. The 2C+, 3C+ and 4C+ constructs contain two, three or four tandem copies of the 49-bp sequence, respectively. The graph shows the level of GUS induction while using various constructs.

An ACGT box is necessary but not sufficient for ABA induction.

ACGT is also present in non-ABA responsive promoters. This suggests that the flanking region of ACGT or some other cis-acting element must be confering the specificity to ABA response. To investigate the specificity, Shen and Ho (1995) analysed the barley HVA22 and HVA1 promoters using loss and gain function experiments. They fused GUS gene with the complete as well as a range of truncated HVA22 or HVA1 promoter.

Shen *et al.* (1996) introduced mutations in the 49-bp fragment (HVA22) and checked for GUS expression. They found that when the last 9 bp (that did not include ACGT box) of the 49 bp promoter fragment is replaced with a random sequence, the expression drops to 44-fold. This result clearly indicated that in addition to an ACGT box, the last 9 bp were also necessary for ABA induction. This 9-bp represents a novel *cis*-acting sequence involved in the ABA response. This was designated as CE1 (coupling element 1) (Figure 9.18).

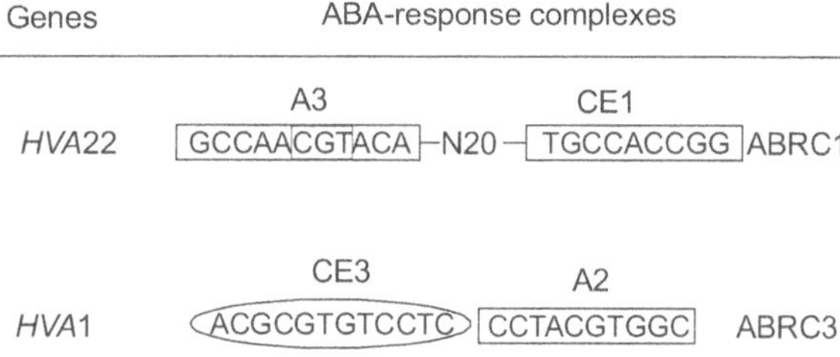

Figure 9.18 The modular nature of ABA-responsive barley genes, *HVA1* and *HVA22*. In *HVA22*, an ABA-responsive complex is composed of an ACGT box, A3 and a distal coupling element (CE1). In *HVA1*, in contrast, an ABA-response complex consists of an ACGT box, A2 and a proximal coupling element (CE3).

A similar approach was used to study the 68-bp HVA1 promoter fragment. Four regions in this sequence were replaced by random sequences designated as fragments I to IV. Most drastic mutations were in fragment region III and IV. Fragment IV had displaced the ACGT box and hence the inhibition in GUS expression was expected. Fragment III replacement caused a reduction in GUS expression. This region again included a *cis*-acting element that did not have homology with any other *cis*-acting elements including CE1. This was designated as CE3.

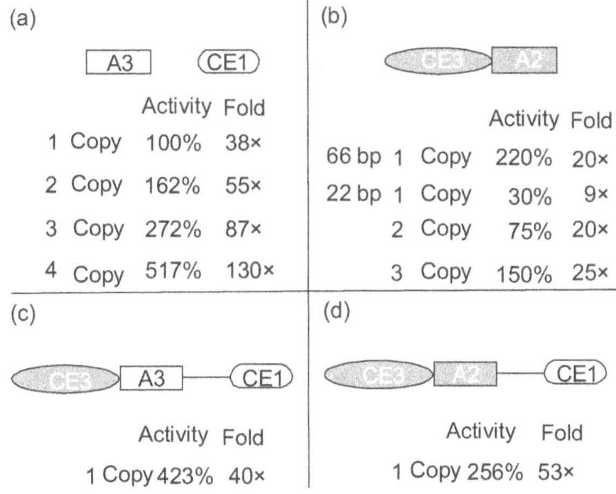

Figure 9.19 Versions of DNA molecular switches controlling the expression of ABA-inducible promoters. (a) HVA22 complex consists of an ACGT box (A3) and a distal CE1. The normal GUS activity from the ABA-treated sample of the single copy *ABRC1* construct is taken as 100% throughout the figure. 'Fold stands for fold induction calculated as described (Shen *et al.*, 1996). (b) The ABA-response complex in HVA1 promoter consists of an ACGT box (A2) and the proximal CE3. (c) and (d) The ternary ABA-responsive complexes consisting of two coupling elements and an ACGT box.

An ACGT box could interact with either a distal or a proximal coupling element (CE) to confer ABA response (Figure 9.18). It has been shown that to achieve a high level of ABA induction, the ACGT box sequence in the HVA22 promoter, i.e., A3 has to interact with a distal element, the 9-bp CE1. The ACGT box in *HVA1* is designated as A2 and this interacts with the proximal element, CE3. An exchanging experiment (exchange of A2 and A3) demonstrated that the two ACGT boxes are fully exchangeable (Shen *et al.*, 1996). Hence, it appears that an ACGT box can interact with either the distal coupling element (CE1) or a proximal element (CE3) to form a promoter complex capable of conferring a high level of ABA induction.

Thus, ABA response relies on the interaction of an ACGT box and a coupling element. Two ABA-response complexes have been classified. They are *ABRC1* and *ABRC3*.

The orientation of both elements is critical for ABA induction in *ABRC1* while the orientation of the *ABRC3* elements has less impact on the ABA response of the complex (Figure 9.19).

Level of induction of ABA-response also depends on the distance between ACGT box and the coupling element (Table 9.1).

Table 9.1   Effect of the number of bases between the ACGT box and CE1 on the level of GUS induction upon addition of ABA

No. of bases between ACGT and CE in ABRC1	Level of induction in response to ABA
5 bp	5-fold
10 bp	16-fold
20 bp	20-fold
30 bp	26-fold

Thus ABA induction level **increases**, as the distance between the ACGT box and the coupling element is lengthened.

In contrast, the induction level in ABRC3 **decreases** as the distance between A2 and CE3 is lengthened (Table 9.2).

Table 9.2   Effect of the number of bases between the ACGT box and CE 3 on the level of GUS induction upon addition of ABA

No. of bases between ACGT and CE in ABRC3	Level of induction
5 bp	26-fold
20 bp	19-fold
25 bp	6-fold

By analysing mutants, the following boxes have been found necessary for various responses (Figure 9.20).

ABA molecular switches are not only functional in the aleurone layer of barley seeds but also in the vegetative tissues of the barley seedlings.

ABA-responsive promoter switches are regulated by ABA, water deficit and NaCl treatment in stably transformed rice plants.

The following experiment was used to test whether the ABRC1 that is functional in transient assays is also functional in stably transformed plants.

One to 4 copies of ABRC1 were fused to the minimal (–100) promoter of the rice actin (*Act*1) promoter, which was linked to the coding region of the GUS reporter gene (Figure 9.21).

This synthetic construct was introduced into rice embryos. From the embryos, transgenic plants containing one or multiple copies of the transgene were generated. The ABRC1 in these transgenic plants was responsive to the following treatments.

1. 50 μ ABA for 20h

2. drought stress (without water for 6 days)

3. salt stress (150 mM NaCl for 72 hours)

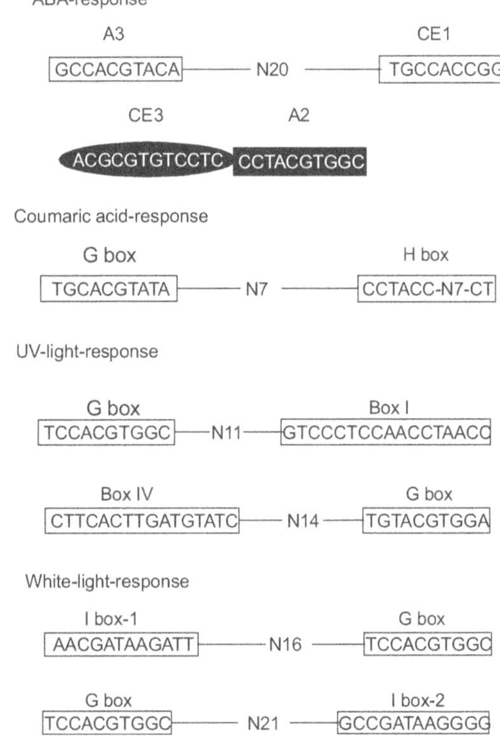

Figure 9.20 Schematic model of signal-specific complexes N-nucleotides

Figure 9.21 Synthetic construct used in rice embryos

Constructs containing 4 copies of ABRC1 confer higher levels of ABA/stress induction than the construct containing only a single copy of ABRC1.

ABRC1 and ABRC3 are mediated by different signal transduction pathways.

Viviparous 1 or vp1 codes for a transcription factor involved in ABA induction (McCarty *et al.* 1991). The ABA-induced expression of some genes, for instance, maize Rab28 (Pla *et al.*, 1991), is VP1-independent while the induction of others, such as the wheat *Em* gene was found to be VP1-dependent for its expression.

When *vp*1 gene was transformed with either ABRC3 or ABRC1, it resulted in different levels of ABA response. Level of induction was quantitated using the reporter construct (GUS) (Table 9.3).

ABRC3, but not ABRC1, is activated by the maize VP1 transcription regulator. This is clear from the following data of Shen *et al.*, 1996.

Table 9.3   Role of VP1 in GUS inducibility upon treatment with ABA

	Presence or absence the VP1	ABA treatment	Fold induction of ABA response (as measured by GUS activity)
ABRC1	–	–	–
	+	–	1 ×
	–	+	15 ×
	+	+	17 ×
ABRC3	–	–	–
	+	–	4 ×
	–	+	14 ×
	+	+	31 ×

POTENTIAL USE OF HORMONE-RESPONSIVE ELEMENTS TO CONTROL GENE EXPRESSION IN PLANTS

Hormone-responsive elements (HRE) are minimal DNA sequence motifs that confer hormone responsiveness to a promoter. Over the last few years, a number of plant hormone-responsive promoters have been studied. There are several different types of auxin-response elements (AuxREs), including the octopine synthase (ocs) or activator sequence-1 (as-1) element, natural composite AuxREs containing TGTCTC elements, synthetic composite AuxREs, and synthetic TFTCTC AuxREs that function without a coupling element. *trans-* factors that interact with these AuxREs have also been identified.

OCS/AS-1 AuxRE

The ocs element was originally identified as an enhancer element in the promoter of *Agrobacterium tumefaciens* octopine synthase gene which is transferred to plant cells via the T-DNA (Ellis *et al.*, 1987). Functional ocs elements were subsequently identified in other opine synthase genes (e.g. nopaline synthase and mannopine synthase) from *A. tumefaciens* T-DNAs and in the −75 region of the cauliflower mosaic virus (CaMV) 35S promoter. In CaMV, this element is referred to as ocs/as-1. The ocs/as-1 element, in opine synthase gene and DNA-virus promoters, has been reported to be activated by exogenous auxin application to plant cells and tissues (Kim *et al.*, 1994).

The ocs/as-1 element is a 20-bp DNA sequence that consists of a direct repeat separated by 4 bp, with the consensus sequence TGACGTAAGCGCTGACGTAA. Functional ocs/as-1 elements do not generally contain a perfect consensus sequence, but instead contain variations of this element with exact spacing (i.e., 4 bp) separating the direct repeats. The spacing between the direct repeats has been shown to be crucial for ocs/as-1 element activity (Singh *et al.*, 1989). The ocs/as-1 element contains tandem binding sites for transcription factors that are functionally identical and both binding sites must be occupied for ocs/as-1 element activity (Singh et al., 1989). Mutations in one of the two binding sites result in loss of ocs/as-1 element function. DNA-binding proteins that interact with the ocs/as-1 element have been identified in plant nuclear extracts. Transcription factors that bind the ocs/as-1 element have been cloned from a variety of plants.

Gene	Ocs/As-1 element
Consensus	TGACGTAAGCGCTGACGTAA
Ocs (−194 to −175)	AaACGTAAGCGCTtACGTAc
nos (−130 to −111)	TGAgcTAAGCaCatACGTcA
AS-1 (−82 to −63)	TGACGTAAGgGaTGACGcAC

Ocs/as-1 elements from CaMV, and opine synthase have been shown to be responsive to exogenous applications of auxins, salicyclic acid (SA), and/or methyljasmonic acid (mJA) (Kim *et al.*, 1993). The ocs/as-1 elements from the soybean Gmhsp 26-A and tobacco *NT*103 genes have been reported to be equally responsive to both biologically active auxins [e.g. indole acetic acid (IAA), α-napthaleneacetic acid (α-NAA), 2,4-dichlorophenoxy acetic acid (2,4-D), 2,4,5-trichlorophenoxy acetic acid (2,4,5-T) and biologically inactive or weak auxin analogues (e.g. 2,3-D, 2,4,6-T, α-NAA).

Natural Composite AuxREs

Auxin-responsive genes, such as soybean GH3, SAURs, Aux22 and Aux28 and pea PSIAA4/5 and PSIAA6, are activated specifically by biologically active auxins and not by other agents (Hagen *et al.*, 1984). The promoters in these genes contain no apparent ocs/as-1 element. Instead, these promoters contain one or more copies of a conserved element, TGTCTC, or some variation of this element (e.g. TGTCCC, TGTCAC) within small promoter-regions that confer auxin responsiveness.

Fine-structure mapping of the AuxREs in the soybean GH3 promoter revealed that TGTCTC elements were critical for AuxRE function (Liu *et al.*, 1994) The GH3 promoter contains three AuxREs that can function independently of one another, and each of these AuxREs, designated E1, D1 and D4, contributed to the overall activity and auxin-inducibility of the GH3 promoter and to the tissue- and organ-specific expression patterns of the GH3 gene. A diagram of the AuxREs in the GH3 promoter is shown in Figure 9.22.

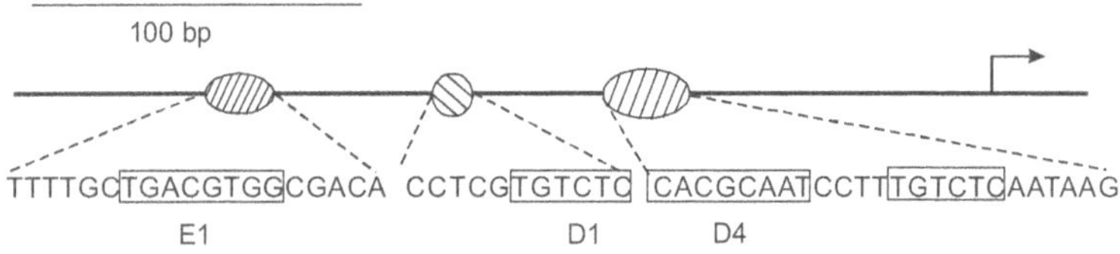

Figure 9.22 Diagram of the soybean GH3 promoter and its composite AuxREs. A 300 bp GH3 promoter is shown with relative positions of three AuxREs: E1; D1; and D4. The D1 and D4 composite AuxREs contain a TGTCTC element (arrows). The E1 AuxRE contains a TGA box or G box that overlaps with a TGTCNC element (arrow showing inverse orientation). The transcription start site is indicated by an arrow at the top. Boxed sequences include TGA box or G box in E1 and functionally defined constitutive or coupling elements in D1 and D4.

Ulmasov *et al.* (1995b) showed that the sequence TGTCTC in the D1 and D4 AuxREs of the AuxRES of the GH3 promoter confers auxin responsiveness when coupled to a closely associated constitutive element. These coupled elements are referred to as composite AuxREs. Composite AuxREs are defined as two adjacent or overlapping elements, a constitutive element and TGTCTC element that work in combination to confer auxin responsiveness to a promoter. A constitutive element is defined as an element that confers constitutive expression to a minimal promoter (i.e., –46 CaMV 35S RNA promoter)-GUS reporter gene in transfected protoplasts.

The composite nature of the D1 and D4 AuxREs in the GH3 promoter is shown in Figure 9.23. The D1 composite AuxRE is an 11-bp element (Ulmasov et al., 1995a). A multimerized D1 construct (D1-4) fused to a minimal promoter-GUS reporter gene is induced about sixfold by auxin in transient assays with carrot protoplast (Table 9.4). Mutations in the 3′ half of the TGTCTC (D1-3 and D1-5 constructs) result in loss of auxin responsiveness and a gain in constitutive expression, while a mutation 5′ to the TGTCTC element (D1-6) results in loss of promoter activity. These results along with other results (Ulmasov *et al.*, 1994) indicate that in the D1 AuxRE, the TGTCTC represses constitutive expression and is required but not sufficient for auxin responsiveness. Constitutive expression is conferred by sequence upstream and including the 5′ region of the TGTCTC element (compare D1-1 construct with D1-3 and D1-5). Thus, the D1 AuxRE contains a constitutive element that overlaps with TGTCTC element.

Table 9.4 Composite nature of D1 and D4 AuxREs

D1 series	Composite AuxRE	GUS units –/+α-NAA	Fold auxin induction
D1-4 (4X)	CCTCGTGTCTC	155/950	6.1
D1-3 (4X)	CCTCTGTaaa	390/420	1.1
D1-5 (4X)	CCTCGTGgaTC	430/450	1.0
D1-6 (4X)	AaaaaTGTCTC	55/55	1.0
D1-1 (4X)	CCTCGaaaaaac	30/30	1.0

Synthetic Composite AuxREs

Ulmasaov *et al.* (1995b) constructed two synthetic composite AuxREs to determine if the TGTCTC element could confer auxin responsiveness when paired with constitutive elements that were not resident in natural composite AuxREs. One of these constructs consisted of chicken c-Rel DNA-binding sites fused or not fused to TGTCTC. In the absence of the TGTCTC element, the c-Rel construct conferred constitutive expression to a minimal –46 CaMV 35S RNA promoter-GUS reporter gene when transfected in carrot suspension culture protoplasts, indicating that these protoplasts possessed an endogenous transcriptional activator that recognized the cRel-DNA-binding sites. This constitutive expression was repressed when a TGTCTC element was placed immediately downstream of the c-Rel DNA-binding sites. Addition of auxin relieved this repression and resulted in activation (i.e., 12-fold auxin inducible).

Synthetic Simple AuxREs

While natural AuxREs within promoters of genes that specifically respond to auxin may generally consist of composite elements, the question remained open whether TGTCTC and related elements might function as AuxREs in the absence of a coupling element if multimerized and properly organized. To test this possibility, Ulmasov *et al.* (1997a, b) constructed a number of direct tandem repeats and palindromic repeats of the TGTCTC element. A D1–4m construct was created by carrying out site-directed mutations in the constitutive portion of the natural 11-bp D1–4 composite AuxRE (Table 9.5). The multimerized (i.e., six- or seven-tandem direct repeats of 11 bp) D1-4m synthetic AuxRE was found to have several-fold greater activity and auxin inducibility (i.e., 29- to 50-fold) in transient assays with carrot protoplasts than the multimerized, natural D1-4 AuxRE (i.e., sixfold auxin inducible).

Table 9.5 Simple TGTCTC AuxREs

D series	Simple AuxRE	GUS units -/+α-NAA	Fold auxin induction
D1-4 (4X)	CCTCGTGTCTC	155/950	6.1
D1-4m (6X)	CCTttTGTCTC	60/1729	29.0
D1-4m (7X)	CCTttTGTCTC	33/1707	52.0

Using Hormone–Responsive Promoters to Control Gene Expression

Studies on the AuxREs discussed above, as well as studies on other plant HREs, suggest possible strategies for controlling gene expression with hormones and/or other chemical agents. Novel approaches to regulate the expression of selected genes might be achieved by taking advantage of the promoters containing the ocs/as-1 element or composite AuxREs and other composite plant HREs. Ocs/as-1 elements fused to minimal promoter-GUS reporter genes have been reported to be induced by most of the biologically active auxins and inactive auxin analogues.

In combination with other *cis*-elements, the ocs/as-1 element might be induced by either a wider or narrower spectrum of hormones and non-hormonal chemical agents. By building synthetic promoters containing specific *cis*-elements along with the ocs/as-1 element, it might be possible to achieve specific particular expression patterns and the inducibility desired.

REVIEW QUESTIONS

1. Explain inducible control of gene expression in plants with examples.

2. Give a detailed account on the mechanism and regulation of *Tn*10 encoded resistance.

3. How is the ecdysteroid responsive expression system used in plants?

4. What are the structural features of a nuclear receptor

5. Explain in detail the basis and functioning of the copper-controllable expression system.

6. Discuss tissue-specific gene expression in plants.

7. Describe the role of chemically induced systems in the control of gene expressions in plants.

8. How will you identify the expression of transgene in higher plants?

9. Give an account of the factors that mediate nitrate-induced gene expression.

10. What are the applications of heat-shock promoters?

11. Describe the regulatory elements present in upstream promoter region of selected abiotic stress-responsive genes in plants.

12. Explain in detail, wound-inducible genes in plants.

13. How can we use the senescence-specific promoter to improve flowering and ripening of fruits?

14. Write an essay on the gene regulatory events associated with senescence.

15. Give a detailed account of the ABA-responsive genes.

16. What are the potential uses of the hormone-responsive elements?

17. Describe the hormone-responsive gene expression in plants.

PART III

APPLIED PLANT BIOTECHNOLOGY

Plant Tissue Culture

Transgenic Plants

Molecular Pharming

The most important application of plant tissue culture is micropropagation. Small amounts of tissue can be used to raise hundreds or thousands of plants in a continuous process. The advantage of using this method is that, about four million genetically identical plants could be obtained from a single apical bud. Plant tissue culture is used for many purposes such as production of secondary metabolites, pathogen-free plants, germplasm conservation and genetic manipulations. By genetic engineering, a number of transgenic plants with improved traits namely pest or disease resistance, better yields, tolerance to heat, cold, salinity and drought, enhanced flowering and fruiting have been produced.

Production of vaccines in plants is one of the best contributions of biotechnology to mankind. Plant-derived vaccines against hepatitis B, cholera, tuberculosis and melanoma are considered safer than the traditional vaccines due to the absence of the contaminating pathogens such as mammalian viruses, prions and toxins.

PLANT TISSUE CULTURE

Establishment of plant calli (groups of undifferentiated cells, singular-callus) and cell suspension culture require optimum culture medium components, proper explant source and plant growth regulator concentrations. A large number of plant species from a wide range of families have now been successfully grown in culture. The following are the major requirements of plant tissue culture.

CULTURE MEDIA

A large number of different culture media are described in the literature, but of these, only a few have found widespread usage for a range of plant species. The commonly used media are Murashige and Skoog (MS) medium and B5 medium.

The six major components of these media are as follows:

1. Major inorganic nutrients
2. Trace elements
3. Iron source
4. Vitamins
5. Carbon source
6. Plant growth regulators

In both MS and B5 media sucrose is the carbon source.

PLANT GROWTH REGULATORS

Most plant culture media contain an auxin and a cytokinin. They are used at a concentration of about 10 mM to maintain cell growth and to promote cell division. 2,4-dichlorophenoxy acetic acid (2,4-D) is the most widely used synthetic auxin. 2,4-D and naphthalene acetic acid (NAA) have replaced the naturally occurring auxin, indole acetic acid (IAA), in tissue culture media.

Benzylaminopurine (BAP) and kinetin are the most commonly used cytokinins. Usually manipulation of auxin and cytokinin levels will be successful in defining a growth regulator balance necessary for the required behaviour in culture.

PREPARATION AND STERILIZATION OF EXPLANTS

A wide range of plant organs and tissues can be used as sources of explants (any part of the plant either meristem or nonmeristem of existing plants that can be induced to form calli or shoots on a medium) for the initiation of callus cultures. The seedlings from sterilized seeds can be used as a source of sterile root, shoot and leaf material. Small seeds can be directly planted on callus induction medium. The inoculated seed will germinate and then the tissues will form the callus directly.

INITIATION OF CALLUS AND SUSPENSION CULTURES

Explants are usually placed horizontally on the surface of a suitable agar-solidified medium taken in a Petri plate. Stem sections may produce more calli if placed vertically with one cut end in agar. Inclusion of an auxin and cytokinin are necessary for callus growth. Higher auxin concentrations are required for callus initiation. The Petri plates are then incubated at approximately 25°C preferably in the dark. Once calli are well established, tiny clumps of calli from actively growing regions are transferred to a fresh medium either for callus propagation or for shoot induction.

Suspension cultures form readily after transfer of callus to flasks containing culture medium minus agar. When the culture maintained in an orbital shaker [30–150 revolutions per minute (rpm)] reaches a suitable cell density, it is subcultured. Callus and cell cultures having the potential to produce plantlets via somatic embryogenesis are referred to as **embryogenic cultures**. Embryogenic cultures are widely used for obtaining transgenic plants.

Growth of callus and cell suspension cultures can be monitored either by an increase in fresh or dry weight or by an increase in cell number. For cell suspension cultures, increase in packed cell volume (PCV) is also a good indicator of growth.

The following are the parameters for obtaining data on cultures.

(a) *Fresh and dry weight measurements* For callus culture, the entire callus (scraped off the medium) is transferred to a pre-weighed weighing boat (or container) to determine fresh weight. For cell suspension cultures, the entire content of the cell culture flask is transferred to a pre-weighed centrifuge tube. The tube is spun for 10 minutes at 200 g. Then the supernatant is pipetted out without disturbing the pellet. The centrifuge tube with cells is weighed to determine the fresh weight. After measuring the fresh weight, the sample is dried in an oven at 60°C until no change in dry weight is observed.

(b) *Increase in cell number* A haemocytometer can be used for determining the cell numbers in a fine suspension culture.

(c) *Packed cell volume* The entire contents of the flask are transferred to a graduated centrifuge tube. It is spun at

200 g for 5–10 minutes and the volume of the pellet is determined. Packed cell volume is expressed as a percentage of the volume of the pellet to the entire culture volume.

ISOLATION, CULTURE AND REGENERATION OF PROTOPLASTS

Protoplasts are cells without cell wall, which provide the starting point for many of the techniques of genetic manipulation of plants. Such experiments consist of three stages.

1. protoplast isolation

2. the genetic manipulation event involving protoplast fusion or gene uptake

3. protoplast culture and regeneration of fertile plants

Protoplasts have been isolated by mechanical disruption or by enzymatic degradation of their surrounding cell walls. Leaf tissues are frequently employed as source materials for protoplast isolation. Young tissues release protoplasts with the highest viability.

Protoplast Isolation (Enzyme Treatment)

The optimum conditions and enzyme treatments for a particular plant genotype/explant combination have been determined empirically. Protoplast isolations are performed at a temperature of 25–28°C for a period of either 2–6 hours or 12–20 hours (overnight).

Following enzyme treatment, a mixture is obtained which contains the released protoplasts, undigested cells and cellular debris suspended in the enzyme solution. This solution is passed through nylon and metal sieves of decreasing pore size to remove the undigested clumps. It is then centrifuged to obtain the pellet containing protoplasts. The protoplasts are resuspended in a washing medium and centrifuged again to remove cellular debris. Centrifugation and resuspension are repeated until a purified protoplast preparation is obtained.

Protoplasts of most species are inoculated at a density between 5.0×10^2 and 1.0×10^6/ml in a semi-solid medium prepared using either agar or agarose.

Protoplast Fusion

This can be mediated by either chemical or electrical techniques. In both cases, the plasma membranes are temporarily destabilized, resulting in pore formation and cytoplasmic linkage between adjacent protoplasts. Polyethylene glycol (PEG), dextran and polyvinyl alcohol (PVA) have been used to induce protoplast fusion, whereas, an equipment consisting of a generator (to produce an AC field), a DC source, a switching unit (to apply the AC field or DC pulses) and a suitable fusion chamber is used in the electrical method of protoplast fusion.

Regeneration of Protoplasts

Sixty days after protoplast isolation, the protoplast-derived tissues are transferred to MS medium supplemented with 1-naphthalene acetic acid (NAA, a synthetic analogue of indole acetic acid, IAA), 6-benzylaminopurine (BAP, which induces rapid multiplication of shoots and agar).

As shoot buds develop, they are detached and transferred to MS salt-based multiplication medium. The rooted plants are finally transferred to pots containing soil-less compost in a glasshouse.

Plant regeneration via embryogenic suspension cultures Embryogenesis is the process of embryo initiation and development. For zygotic embryos, embryogenesis starts at zygote formation, ends at seed maturation. During somatic embryogenesis, an embryo, similar to the zygotic embryo that contains both shoot and root primordia, is formed from somatic plant tissue. Embryonic suspension cultures form aggregates of embryonic tissue. The cells that make up the aggregates of embryogenic tissue are typically small (~20 µm in diameter), isodiametric and densely cytoplasmic with small vacuoles.

Proliferative cultures contain globular and later-staged embryos. Establishment of embryogenic suspension cultures may take from 1–12 months. For further development of embryos, the proliferative embryogenic tissue is transferred to a medium containing either no or low auxin and sometimes ABA, high sugar or amino acids. Somatic embryos germinate to form roots and shoots once they reach physiological maturity.

Somatic hybridization provides a method for generating hybrids between sexually incompatible plants and also facilitates genetic modification of vegetatively propagated crops.

MICROPROPAGATION

Micropropagation is the practice of rapidly multiplying stock plant material to produce a large number of progeny plants, using modern plant tissue culture methods.

Micropropagation is used to multiply novel plants, such as those that have been genetically modified or bred through conventional plant breeding methods. It is also used to provide a sufficient number of plantlets for planting from a stock plant which does not produce seeds, or does not respond well to vegetative propagation.

Micropropagation involves five stages:

(a) **Stage 0:** *The preparative stage* The plant material for *in vitro* culture is prepared. The aim is to obtain hygienic and physiologically active starting material.

(b) **Stage 1:** *Initiation of cultures*   The aim is to obtain a reliable start. Multiplication is not important at this stage.

(c) **Stage 2:** *Shoot multiplication*   The only function of this stage is to increase and maintain the stock. Meristematic centres are induced and developed into buds and/or shoots.

(d) **Stage 3:** *Shoot elongation, root induction and root development*   Shoots are elongated and rooted.

(e) **Stage 4:** *Transfer to greenhouse conditions*   Most species require acclimatization in order to ensure their survival.

For most micropropagation work, the choice of explant is an apical or axillary bud. For a limited number of plants, other explants such as leaf pieces are used. Explants are generally rinsed in 95% ethanol for a few seconds and sterilized for 3–5 min. using 0.1 to 1% mercuric chloride ($HgCl_2$) plus detergent (two drops/100 ml). They are rinsed with autoclaved water several times and placed on media.

Organ differentiation in plants is regulated by interplay of auxins and cytokinins. A higher cytokinin-to-auxin ratio promotes shoot formation and a higher auxin-to-cytokinin ratio favours root differentiation. However in a number of cases, cytokinin alone is enough for optimal shoot multiplication.

Benzylaminopurine (BAP) is probably the most useful and reliable cytokinin. Generally 1–2 mg cytokinin/L is adequate for most systems. High levels tend to induce adventitious bud formation. Since IAA is the least stable auxin in the medium, synthetic auxins such as NAA and IBA (indole 3-butyric acid) are preferred. For shoot multiplication, their concentrations range from 0.1 to 1.0 mg/L. 2,4-D is avoided as it has a strong tendency to induce callus.

On the same medium, stage 2 cultures originally yielding axillary shoots can produce abundant adventitious shoots after a number of subcultures. An unlimited number of subcultures increases the occurrence of variations and mutations. Generally, 10–12 subcultures is considered to be maximum.

High levels of cytokinins in stage 2 produce a high number of shoots. Good elongation of the shoots can be obtained by transferring them to a medium devoid of cytokinin. Adventitious and axillary shoots that develop in the presence of a cytokinin, usually lack roots. It is often very difficult to develop well functioning roots *in vitro*. Roots generally start developing when they are transferred to *ex vivo* conditions. Auxins can be exogenously applied to induce root formation. For most species, auxins such as NAA or IBA are required to induce rooting.

During hardening the culture vessels are uncapped and placed in a greenhouse several days prior to removal of the plants from the culture medium. After washing the agar thoroughly from the plants, they are planted on unfertilized peat, maintaining a high relative humidity.

TRANSFORMATION

The bacterial tumour-inducing (Ti) plasmid of *Agrobacterium tumefaciens* is the most widely used plasmid vector in plant molecular genetics, which aids in the development of modified plants with the help of recombinant DNA technology.

A. tumefaciens is a phytopathogenic, gram-negative, non-sporing, motile bacillus. It is found in the rhizosphere region. The four species of *Agrobacterium* are,

- ✖   *A. tumefaciens*—causes crown gall disease
- ✖   *A. rubi*—causes gall type tumours on stem
- ✖   *A. rhizogenes*—causes hairy root disease
- ✖   *A. radiobacter*—an avirulent species

Crown gall (Figure 10.1) formation occurs through wound infection. It causes cancerous proliferation of the stem tissue in the crown region.

Figure 10.1   Crown gall formed by *Agrobacterium tumefaciens*

Features of Ti Plasmid

The structure of Ti plasmid is given in Figure 10.2. The following are the important features of the Ti plasmid.

✗   Responsible for crown gall disease.

✗   >200 kb

✗   Part of the Ti plasmid-T-DNA (15 to 30 kb) integrates into the plant chromosomal DNA.

✗   Maintained in the stable form in the plant cell.

✗   T-DNA region has 8 genes.

✗   Help in the synthesis of opines—derivatives of arginine

  ✗   Left : TL-T-DNA

  ✗   Right : TR-T-DNA

  ✗   Opines —C and N_2 source for growth and multiplication.

  ✗   Conjugative plasmid

Ti plasmid genes are involved in plant and bacterial functioning.

Locus	Function	Ti plasmid
vir	DNA transfer into plant	All
shi	Shoot induction	All
roi	Root induction	All
nos	Nopaline synthesis	Nopaline plasmid
noc	Nopaline catabolism	Nopaline plasmid
ocs	Octapine synthesis	Octapine plasmid
occ	Octapine catabolism	Octapine plasmid
ori V	Origin of replication	All

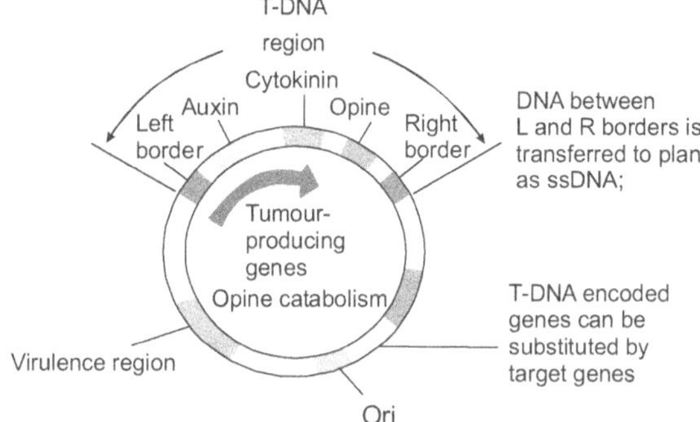

Figure 10.2   Ti plasmid

Sequence of Infection Process

The sequence of the infection process is given diagrammatically in Figure 10.3.

- ✖ *A. tumefaciens* is attracted towards wound site by chemotaxis.

- ✖ Wound induces the production of phenolic compounds especially acetosyringone (10^{-7} M).

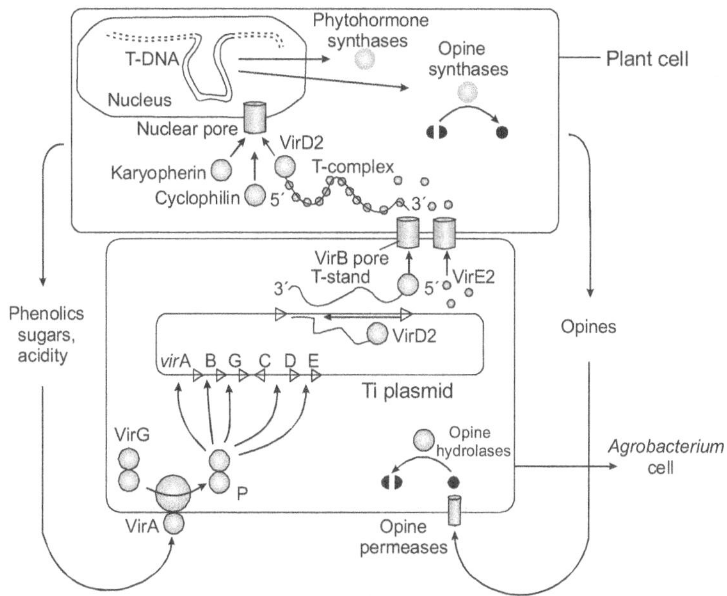

Figure 10.3   T-DNA transfer and its integration into host plant cell genome

- ✗ Lipopolysaccharides aid in attachment.
- ✗ Ti plasmid codes for chemotactic receptors.
- ✗ Acetosyringone (10^{-5} to 10^{-4} M) activates *vir* genes.
- ✗ *vir* genes coordinate the infection process.
- ✗ Production of permeases by plant cells.
- ✗ Phenolic compounds bind to *virA*.
- ✗ *virA* activates *virG*.
- ✗ VirG activates *virD* and *E*.
- ✗ VirD protein is an endonuclease that excises T-DNA.
- ✗ T-DNA is transferred as DNA protein complex—T-complex.

✘ Proteins associated are VirE2-SSB (single-strand binding proteins).

✘ Integrates with plant DNA.

✘ TL-T-DNA encodes two enzymes for synthesis of hormones.

✘ Disorganized proliferation of cells in the plant.

✘ Loci in TL-T-DNA control morphology of galls.

  ✘ Tml : causes large tumours

  ✘ Tmr : induces tumour in roots

  ✘ Tms : induces tumour in shoots

This natural phenomenon is exploited in introducing foreign genes into desired plants. Plants thus transformed with foreign genes are called **transgenic plants**.

Transformation refers to the generation of transgenic organisms, i.e., the introduction of a gene of choice into the genome of an organism. This is also referred to as **genetic engineering** or **genetic modification**.

Transformation of plants requires

✘ *Agrobacterium* as a vehicle

✘ Engineered Ti plasmid

✘ T-DNA with appropriate deletions and inserts

The transformation process involves

✘ Incubation of *Agrobacterium* with plant protoplasts

✘ Killing of bacteria with antibiotic

✘ Tissue culture

✘ Killing of non-transformed cells by kanamycin

✘ Regeneration of plants from transformed cells

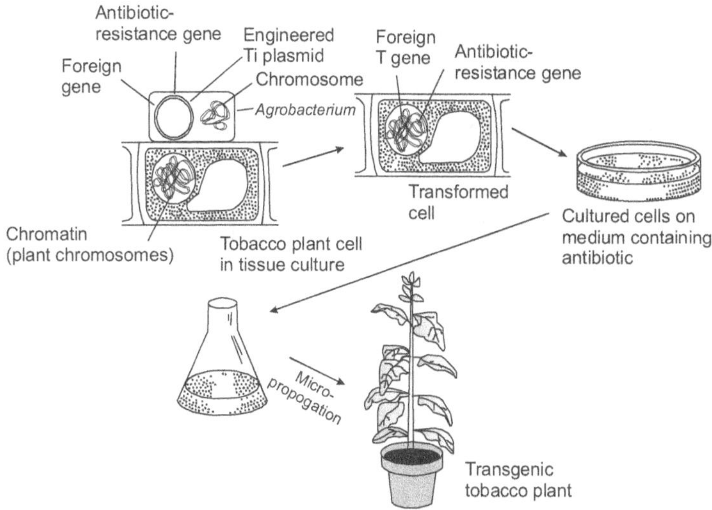

Figure 10.4   Introduction of a foreign gene into plants

The reasons for generating transgenic plants are

✗   Improvement of crop plants by introduction of new genes

✗   Promoter studies

✗   T-DNA tagging to clone new genes

✗   Finding out the function of a gene by overexpressing it or by reducing its expression

Making a transgenic plant requires five major steps.

1. Gene of interest has to be cloned into an expression vector that contains a ribosome-binding site, a promoter and a terminator.

2. Transformation, which is the introduction of a gene of interest into the plant tissue. Several techniques are available.

3. Regeneration of the transformed tissue–growing a new plant from transformed cells.

4. Selection of transformants.

5. Analysis of transgene expression.

Expression vectors (typically plasmids) can be custom-made or bought commercially. To insert the gene of interest into the vector, standard lab procedures are followed that typically involve the use of *E. coli* as a way to get enough vector DNA.

Typical plant expression vectors contain the following features:

- *E. coli* origin of replication so that the plasmid can be manipulated in *E. coli*.

- A bacterial antibiotic resistance gene, typically ampicillin. This allows for the selection of bacteria that contain the plasmid.

- A plant promoter, often cauliflower mosaic virus (CaMV) 35S rRNA promoter. This is a strong promoter that is expressed in a lot of plant tissues. Other promoters may be more suitable, particularly if the gene of interest needs to be expressed in a tissue-dependent manner.

- A plant selectable marker. This can be a herbicide resistance gene, or an antibiotic resistance gene (often kanamycin).

- For *Agrobacterium* vectors there is a separate origin of replication and there may be a separate antibiotic resistance gene (e.g. hygromycin).

- After construction of the vector using standard techniques (restriction digestion, ligation), the vectors are introduced in *E. coli* and then propagated. The plasmid is then harvested and is either used for direct transformation or is introduced in *A. tumefaciens*.

Direct Transformation

The most common technique for direct transformation is **microprojectile** or **particle bombardment** (Figure 10.5). This technique is based on the use of a "particle gun" or "gene gun". The expression vector with the gene of interest is precipitated onto tungsten or gold particles which are then shot into the plant tissue. In most cases only transient expression (i.e., the DNA does not integrate into the genome but is transcribed until it degrades) is observed. In a small percentage of the cells the DNA will integrate and stable expression is seen. The main drawback of this technique is that often multiple copies of the transgene are inserted.

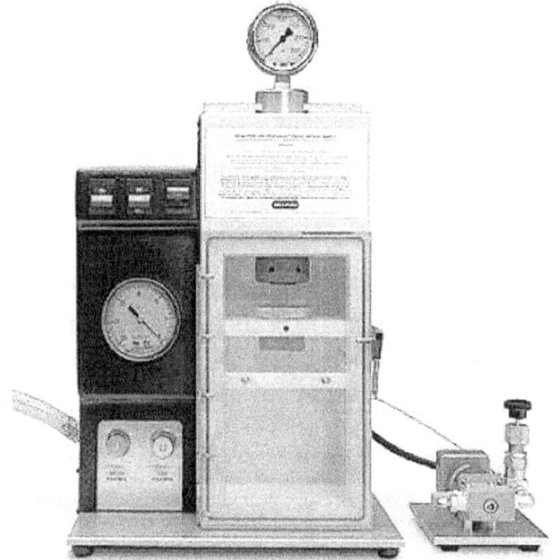

Figure 10.5 Particle bombardment system

Agrobacterium-mediated Transformation

This transformation process involves infecting the plant tissue with *A. tumefaciens* carrying the newly designed Ti plasmid.

The general method involves the use of a leaf punch that is incubated in a suspension of bacteria. The cells on the edge of the leaf punch are transformed. These transformed cells now need to develop into new plants through tissue culture. The leaf disc is typically placed on a medium that stimulates shoots (high cytokinin : auxin ratio), selects for transformants (based on the selectable marker) and kills *Agrobacterium* (carbenicillin). The shoots that form are placed on a root-inducing medium (low cytokinin:auxin ratio) and the resulting plantlets are eventually transferred to soil.

This regeneration process is very similar for tissues transformed with the particle gun.

The latest *Agrobacterium*-based transformation technique in *Arabidopsis* is referred to as "floral dip". In this case the tissue culture stage has been eliminated. Instead, the *Arabidopsis* plant is dipped into a suspension of *Agrobacterium* at the stage where it has the maximum number of unopened floral bud clusters. The suspension also contains a detergent to reduce surface tension, and some sugar. The bacteria get into the developing flowers and transfer the T-DNA. This results in the formation of transgenic seeds that can be planted out on medium with a selectable marker. The plants that survive on the selectable medium get transplanted and can be studied or screened.

Alternative methods of transformation are,

- *Electroporation of plant protoplasts/pollens/embryos/ callus*   In this case the DNA is introduced while the cells are subjected to an electric discharge.

- *Microinjection*   The DNA is injected into the protoplast with a pipette.

- *Silicon-carbide-mediated transformation*   DNA and silicon carbide fibers are mixed with cells and homogenized in a blender. The DNA-coated fibres penetrate the cells via small holes created by the fibres and introduce the DNA.

✖ *Pollen-tube pathway*   DNA can be applied to cut styles shortly after pollination so that it can flow down the pollen tube to end up in the zygote.

✖ *Liposome-mediated transformation*   DNA is loaded into phospholipid spheres (liposomes) and these liposomes are mixed with protoplasts to result in lipofection.

These alternatives are not very efficient compared to *Agrobacterium*-mediated transformation.

Agrobacterium Versus Particle Bombardment

In general, *Agrobacterium* is considered the method of choice for transformation.

Advantages include:

✖ low-copy number of the transgene

✖ higher proportion of stable transformants

✖ larger DNA segments can be transferred

✖ faster method

The selection based on a selectable marker is supposed to result in plants that actually have been transformed, but in some cases only the marker gets transferred, but not the gene of interest. It is possible to check for the presence of the transgene by doing a Southern analysis. Independent transformants are expected to show hybridizing bands of different sizes. It is also possible to check for integration by polymerase chain reaction (PCR).

REVIEW QUESTIONS

1. Explain micropropagation of an explant.

2. Describe the role of protoplast in plant tissue culture.

3. Describe *Agrobacterium*-mediated transformation of gene into a plant cell.

4. Write short notes on direct transformation.

TRANSGENIC PLANTS

INTRODUCTION

Plants are a key source of energy for the survival of mankind. Over the last few decades, plant genomics has been gaining importance for the improvement of plant characteristics. New approaches to GM (genetically modified) crop production are becoming apparent and might help to make the technology more acceptable. These new approaches rely on plant genes to produce the desired effects.

Plant biotechnology began in 1983 when it was demonstrated that foreign genes could be transferred from *Agrobacterium tumefaciens*, a soil bacterium, into plant genome and expressed (Fraley *et al.*, 1983). Today, any gene of interest from any species may be identified, isolated and incorporated across species barriers.

Transgenic plants refer to those genetically modified plants in which functional foreign gene(s) have been inserted in their genome. They contain a gene or genes, which have been artificially inserted instead of the plant acquiring them through pollination. The inserted gene sequence, known as the **transgene**, may come from another unrelated plant, or from a completely different species. These crops are also referred to as **GM** crops.

Transgenic technology enables one to bring together the useful genes from a wide range of living sources in a single plant, not just from within the crop species. This technology provides a means for identifying and isolating the genes controlling specific characteristics and also helps in moving copies of those genes into another organism. This enables generation of plants containing a new combination of genes resulting in more useful and productive plants.

APPLICATIONS OF TRANSGENIC PLANTS

Plant genes controlling or influencing agronomically important traits have been identified and manipulated to produce agronomic benefits. The various applications of transgenic plants in agriculture and other fields include the following.

1. Producing transgenic plants with resistance against

   i. diseases caused by bacteria, fungi, viruses and other microorganisms

   ii. pesticides, herbicides, insecticides and other chemicals

   iii. environmental stresses such as drought, increased alkalinity in the soil, increased salinity in the soil and waterlogging

2. Improving crop yield and quality by producing crops with increased photosynthetic efficiency and harvest index in terms of yield and quality. These include crops which produce material with

   i. longer shelf life, better flavour, etc.

   ii. increased nutritional quality such as higher vitamin and protein content and increased polyunsaturated fatty acid (PUFA) content in case of oil seeds.

3. Developing crops producing value-added feed for animals.

4. Engineering plants to produce prophylactics, vaccines and antibodies (plantibodies) with the added advantage of being taken in orally.

In this chapter, we will discuss a few examples of transgenic plants such as herbicide-resistant plants, stress-tolerant plants and crops with improved yield and quality.

HERBICIDE–RESISTANCE

Herbicide resistance in crop plants is of considerable importance and of potential commercial benefit to chemical companies who produce herbicides. Herbicide-resistant plants have the ability to grow in the presence of herbicide whereas the weeds do not.

Herbicide resistance was one of the first genetically modified (GM) traits to be tested in the field for commercial purposes. In fact, herbicide-resistant and insect-resistant crops are the most widely grown crops, because information is available regarding the modes of action of certain herbicides, and the knowledge of the mechanism involved in the resistance helps in selecting target genes for manipulation.

Herbicides are more toxic to plants than to animals, and the activity of many of these involves a single enzyme/protein as target.

Mullineaux (1992) has identified four distinct strategies based on genetic engineering (GE) to bring about herbicide resistance in crop plants.

Strategies for Genetically Engineering Herbicide Resistance in Plants

1. Over-expression of the target protein (the protein inhibited by the herbicide)

2. Introduction of mutant gene encoding protein with modified sensitivity to the herbicide

3. Detoxification of the herbicide using a foreign gene product

4. Stimulating the plant's own defence against toxic compounds

The herbicides, glyphosate and phosphinothricin, are commercially used and so the discussion of resistance against these herbicides is dealt with, in this chapter. Two case studies, the first one on glyphosate resistance and the second on phosphinothricin are discussed.

Glyphosate Resistance

Glyphosate is a broad-spectrum herbicide claimed to be effective against 76 out of 78 world's worst weeds and can be rapidly degraded in the soil. Monsanto, an American based company, markets this herbicide as **Roundup**. It competitively inhibits the enzyme 5-enol pyruvyl shikimate 3-phosphate synthase (EPSPS), which functions in the biosynthesis (Figure 11.1) of essential aromatic amino acids (tryptophan, tyrosine and phenylalanine). It is a substrate analog of phosphoenol pyruvate—a substrate of the EPSPS enzyme.

Now let us see how the strategies previously are applied in developing glyphosate resistance.

Strategy 1: Overexpression of EPSPS gene Amplification (increase in number of copies of the gene) of EPSPS using a strong promoter in cultured Petunia cells was observed to confer resistance to glyphosate. This resistance was thus due to an increase in the amount of the enzyme. The EPSPS cDNA from petunia was fused to a CaMV 35S (cauliflower mosaic virus 35S RNA) promoter and 3' nos (nopalinae synthase gene) terminator sequence in the vector pMON 546 and transformed into petunia using *Agrobacterium*. A chloroplast transit peptide (CTP) sequence was already present in the EPSP synthase cDNA of petunia, which removed the problems in targeting the protein to the chloroplast.

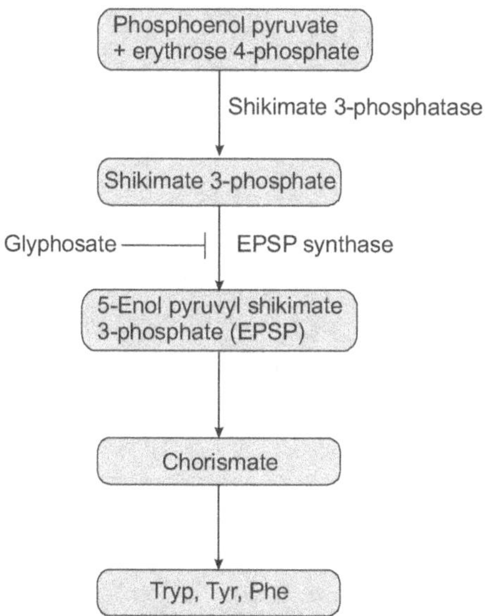

Figure 11.1   Role of EPSPS enzyme in the synthesis of aromatic amino acid

CaMV 35S RNA—promoter sequence from cauliflower mosaic virus; pet CTP—chloroplast transit peptide from petunia pet EPSPS—the target gene to be overexpressed; 3′nos—transcription terminator sequence from the nopaline synthase gene of *Agrobacterium tumefaciens* Ti plasmid.

This results in 40-fold increase in EPSPS activity in the transgenic plants. They were observed to tolerate 2–4 times higher glyphosate concentration than required to kill wild type plants.

Strategy 2: Mutation of the target protein   As we discussed earlier, glyphosate inhibits the action of EPSPS enzyme by binding to it. A mutation in the EPSPS gene could result in a modification in the enzyme, which in turn would reduce its affinity for glyphosate. One such gene was identified in a herbicide-resistant *Agrobacterium tumefaciens* strain CP4. This enzyme product had low affinity for glyphosate and high affinity for phosphoenol pyruvate (its substrate). A construct was created with this gene fused to a CaMV 35S promoter, 3'nos terminator and a chloroplast transit peptide (CTP) from *Arabidopsis*.

The construct used for engineering glyphosate resistance using a mutant EPSPS gene from *A. tumefaciens* is shown in Figure 11.2.

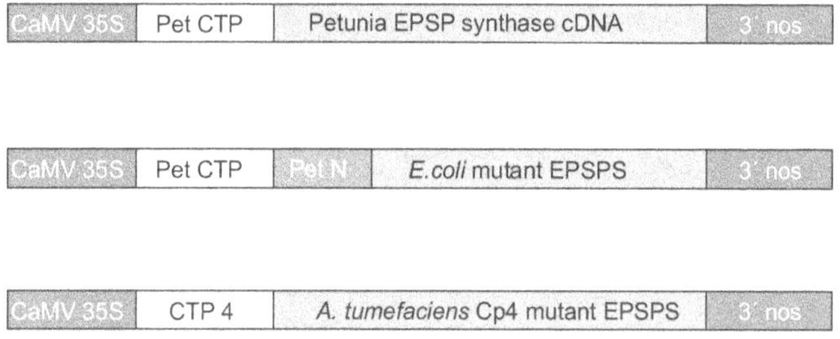

Figure 11.2   The construct used for overexpression of EPSP synthase for engineering glyphosate resistance

These experiments provided evidence for the feasibility of using the mutant protein strategy for developing herbicide resistance.

Strategy 3: Detoxification of the herbicide   Glyphosate is degraded by an oxidoreductase involved in the cleavage of a C—N bond. A gene coding for the enzyme glyphosate oxidase (GOX) performs this function and has been isolated from a soil organism, *Othrobactrum anthropi* strain LBAA, and modified by addition of a transit peptide. Transgenic crops carrying this construct display good glyphosate resistance. Monsanto combined the construct in strategy 2 and 3 in its oilseed rape (canola), i.e., it expressed both the mutant CP4 EPSPS gene and the GOX gene. This approach helped avoiding accumulation of glyphosate by degrading it. Transgenic plants developed by strategy 2 may grow in the presence of glyphosate but may accumulate it more than normal levels.

Strategy 4: Stimulation of detoxification system   This approach is a general one and targets pathways which detoxify xenobiotic compounds in plants. Detoxification usually involves many steps and many enzymes associated with it.

The potential of utilizing the genes encoding enzymes involved in these natural pathways of the plant is yet to be fully exploited.

Phosphinothricin Resistance

Crops resistant to phosphinothricin are next in line to glyphosate in terms of number of crops and acreage of resistant crops. Phosphinothricin (PPT) is another popular herbicide, derived from a natural product, Bialaphos, produced by certain *Streptomyces* species. Bialaphos is a tripeptide of the form PPT-Ala-Ala in which the alanine residues are removed by a peptidase enzyme to form *L*-phosphinothricin—the active form of the herbicide. Phosphinothricin and its derivatives are marketed under different names, one of which is Basta.

Mode of action   Phosphinothricin resembles glutamate in structure and thus it competitively inhibits the enzyme glutamine synthetase (GS)—a key enzyme in nitrogen metabolism. The inhibition of GS causes accumulation of ammonia and inhibition of photosynthesis leading to rapid death of the plant cells.

Strategy for phosphinothricin (PPT) resistance   This is similar to strategy 3 of glyphosate resistance. The *Streptomyces* species that produces bialaphos needs to protect itself from the toxic effects of PPTase and hence has its own detoxification system. The *bar* (bialaphos resistance) gene of *Streptomyces hygroscopicus* and the closely related *pat* gene of *Streptomyces viridochromogens* code for the enzyme phosphinothricin acetyl transferase (PAT). This enzyme adds an acetyl group to the amino group of PPT thus inactivating the compound (Figure 11.3).

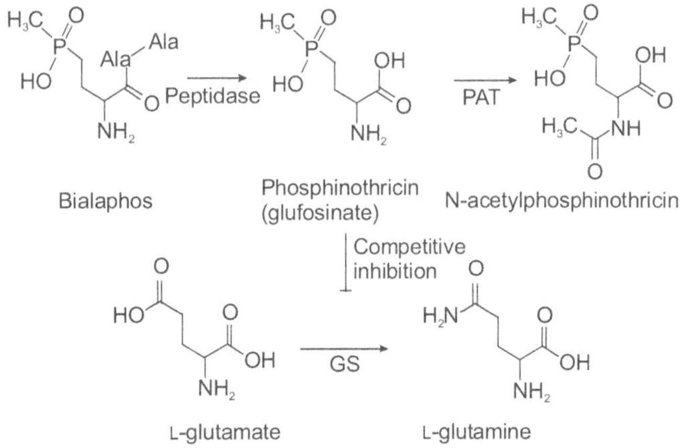

Bialaphos → Peptidase → Phosphinothricin (glufosinate) → PAT → N-acetylphosphinothricin

Competitive inhibition

L-glutamate → GS → L-glutamine

Figure 11.3 Detoxification of PPT by the action of the *pat* gene, conferring resistance to PPT, in the transgenic plant possessing this gene from *Streptomyces* spp.

Limitation

Herbicide resistance and development of "super weeds" The possibility of the resistance genes giving rise to super weeds does exist, though the probability is low. The mechanisms that give rise to super weeds could be one of the following.

1. The resistant crop itself becoming a weed when other crops are planted.

2. Transfer of resistant gene to weeds through pollination.

3. Horizontal gene transfer, e.g. by virus.

In order to enjoy the benefits of the herbicide-resistant plants, issues related to their limitations discussed above have to be addressed.

PEST RESISTANCE

Pests cause loss of about 13% of the potential crop yield. Insect pests cause a major proportion of total pest damage to crops (Table 11.1). Major classes of insects that cause crop damage are the orders Lepidoptera (e.g. butterflies and moths), Diptera (e.g. flies and mosquitoes), Orthoptera (e.g. grasshopper and cricket) and Homoptera (e.g. aphids) and Coleoptera.

Table 11.1 Insect pests affecting the crops

Insect species	Common name of pest	Order	Crops affected
Heliothes virescens	Tobacco budworm	Lepidoptera	Tobacco, cotton
Leptinotarsa decemlineata	Colorado beetle	Coleoptera	Potato
Locusta migratoria	Locust	Orthoptera	Grasses
Nilaparvata lugens	Brown plant hopper	Homoptera	Rice

Engineering Insect Resistance—The *Bacillus thuringiensis* Approach

One of the strategies used for insect resistance is the *Bacillus thuringiensis* approach. The *cry* genes of *B. thuringiensis* represent the best example of using bacterial insecticidal genes for insect resistance. The *cry* gene belongs to a superfamily of related genes. There are many distinct families (*cry* 1–*cry* 40). The protein has three discrete domains.

The first few attempts involved the expression of *cry*1A genes under the control of CaMV 35S promoter in tobacco, tomato and potato plants. But these showed very low expression levels. Then the prokaryotic (bacterial) gene sequence was extensively modified to obtain high levels of stable expression. Eventually, there was 100-fold increase in expression.

Mode of action   The insect larvae feed on the leaves of the transgenic plants expressing the protein in an inactive form. In the midgut, the protein crystals are solubilized and cleaved proteolytically to release the active protein. This causes osmotic lysis of the cells of the epithelium in the midgut proving toxic to the larvae. Different classes of cry proteins have been identified which can be used to effectively kill different insects.

Other Approaches

Some strains of cowpea (*Vigna viniculata*) grown in Africa were found to be resistant to insects and were attributed to the cowpea trypsin inhibitor (*CpT1*). This protein affected only the insect protease, i.e., trypsin. (It had no effect on mammalian trypsin and therefore was not toxic to mammals.) Field trials have shown a significant level of protection of crops against lepidopteran pests and larvae.

Aprotinin is another protease inhibitor that effectively controls the insects belonging to Lepidoptera, Homoptera and Coleoptera.

DISEASE-RESISTANT PLANTS

Plant diseases cause a lot of damage to crops and result in extensive economic loss. The chemicals used to control these diseases are very expensive and further cause problems to the environment and human health. The potential of using genetic manipulation to engineer disease resistance is being explored.

It requires knowledge of the range of diseases affecting plants as well as the natural defence mechanisms of the plant.

The organisms that infect plants could be bacteria, fungi, viruses or phytoplasma (organism similar to mycoplasma that lacks a cell wall) (Table 11.2).

Table 11.2 Pathogens causing serious economic loss

Pathogens	Major crop infected (disease caused)
Fungal and water moulds	
Bipolaris maydis	Maize (Southern corn leaf blight)
Fusarium spp.	Wheat barley (scab) Cotton (wilt)
Phytophthora spp.	Potato (blight) Cocoa (blackpod) Soybeans (rootrots)
Rhizoctonia solani	Many major crops
Viral	
Barley yellow dwarf virus	Major small grain cereals
Rice tungro virus complex	Rice
Tomato spotted wilt virus (TSMV) tospovirus	Peanuts (spotted wilt disease)
Bacterial	
Agrobacterium spp.	Range of dicot plants
Pseudomonas syringae	Beans (blight, brown spot)
Xanthomonas campestris	Tomato pepper beans (bacterial spot)

Plants have various levels of defence against the pathogens attacking them. The first level of defence is the anatomical or structural barrier, for example, cuticle, bark, etc. The second level constitutes the antimicrobial compounds and pre-existing proteins. Secondary metabolites such as terpenes, etc., are examples of chemicals acting as antimicrobials. Then come the inducible and systemic responses of the plants as defence against infection. The genes involved at this level have the potential to engineer resistance against diseases.

Here we will discuss a few strategies to engineer resistance against diseases. Since the plant's defence strategies against fungal and bacterial diseases overlap, the strategies are discussed together. Viral disease resistance will be dealt with separately.

Resistance to Fungi and Bacteria

This involves the genetic manipulation of genes that confer disease resistance in plants.

The 'R' genes 'R' genes or resistance genes are important components of the monitoring system of the plants for pathogen (Figure 11.4).

The *R* gene products of the plants recognize the avirulence (Avr) proteins produced by some pathogens. The presence of *R* gene in the plant and its corresponding *Avr* gene in the pathogen results in disease resistance. If either of them is inactive or absent, the result is disease. In plants, the interaction of Avr protein and the R protein brings about a hypersensitive response (HR) including local necrosis and induction of a cascade of reactions, and confers a degree of resistance against the pathogen. Thus, introduction of the R gene into a plant confers the ability to recognize the pathogen on the plant. For example, in pepper, the B2 gene confers resistance against bacterial spot disease.

Avr genes and elicitor molecules Elicitors are molecules released by the pathogens into the plant. Recognition of these molecules activates a defence response in the plant to inhibit the pathogen. Disease resistance can be engineered by expressing this component of the pathogen in a plant. Some examples of the elicitors are *Avr*-coded proteins, bacterial structural components, toxins, cell wall components such as chitin and melanin, enzymes that degrade plant polymers such as pectate, lyase and cutinase, enzymes that function in overcoming plant defences such as tomatinases, and various secreted peptides like pep131, Avrpita, and elicitins.

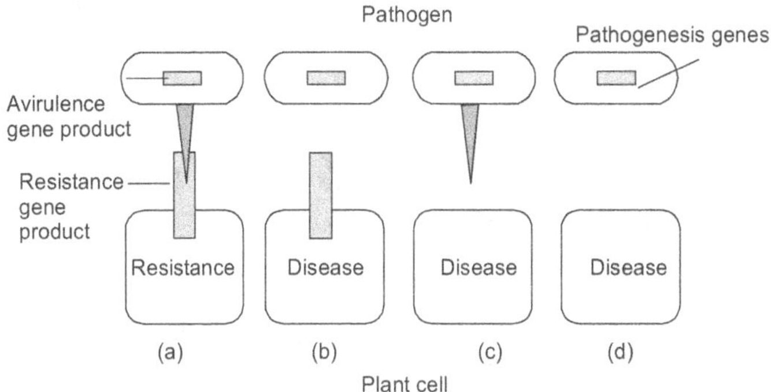

Figure 11.4   HR interactions. In (a) the plant cell contains the resistance gene (*R*) and the pathogen produces the avirulence gene produce (Avr); the plant shows resistance to infection. In (b) the pathogen is not producing the Avr protein so the interaction results in disease. In (c) the host cell is not producing the R protein, so disease results. In (d) neither the R protein nor the Avr protein is being produced, so disease results.

The master switch genes   Pathogen-induced signalling pathways in plants have been found to include the expression of certain master switch genes such as those coding for kinases and transcription factors. Manipulation of these could also lead to increased disease resistance.

These genes act as master-switches by controlling the expression of several genes in a single pathway, thus producing large changes in a single trait, in this case disease resistance with few side effects on other traits, e.g. WRKY transcription factors have been shown to be important in resistance to pathogens such as *Phytophthora infestans*. Though it is difficult to identify best candidate genes from such multigene families, good candidate *WRKY* genes have been identified in *Arabidopsis*. When *WRKY* 18, 29 and 70 were overexpressed in some plants, they showed enhanced resistance to *P. syringae*.

As mentioned earlier, protein kinases such as MAP kinases (mitogen-activated protein kinases) are an integral part of cell signalling cascade reaction and therefore are also good candidates as target for genetic engineering. Other signalling molecules that are nodes in signalling networks are also potential targets for manipulation.

Pathogenesis–related (PR) genes   The pathogen-related elicitors give rise to reaction in the plant cells. There are several parts associated with this reaction. Usually there is damage to plant cell walls causing release of fragments such as pectic oligomers. They act as signals called **endogenous elicitors** that bind specific receptors setting off a cascade of reactions leading to induction of specific defence genes. These genes code for enzymes that synthesize structural components for cell wall thickening, enzymes of secondary metabolism, lectins and many pathogenesis-related proteins.

Though not all of these strategies have succeeded, it is an indicator of the scenario of the future.

Resistance to Viral Infection

Virus infection of plants can result in cell death (necrosis), retarded cell division and growth (hypoplasia) or hyperplasia (excessive cell division or expansion). The plants can show symptoms such as growth retardation, distortion, mosaic patterning of leaves, yellowing and wilting. In agriculture, loss due to viral epidemics cause losses adding up to billions of dollars.

Table 11.3 Examples of coat protein-mediated resistance

Plant	Source of coat protein gene	Virus resistance exhibited to
Alfalfa	AlMV	AlMV
Citrus	CTV	CTV
Papaya	PRSV	PRSV
Peanut	TSWV (N-gene)	TSWV
Potato	PVX	PVY
	PVY	PVY
	PLRV	PLRV
Rice	RTSV	RTSV
	RSV (N-gene)	RSV
	RHBV (N-gene)	RHBV
	RYMV	RYMV
Squash	CMV	CMV
	WMV 2	WMV 2
	ZYMV	ZYMV
Sugar beet	BNYVV	BNYVV
Tobacco	TMV	TMV, PVX, CMV, AIMV
	CMV	CMV
	AIMV	AIMV
Wheat	SBWMV	SBWMV
	BYDV	BYDV

List of virus names can be obtained from the web links: http://image.fs.uidaho.edu/vide

AlMV—Alfalfa mosaic virus, BNYVV—Beet necrotic yellow vein virus, BYDV—Barley yellow dwarf virus, CMV—Cucumber mosaic virus, CTV—Citrus tristeza virus, PLRV—Potato leafroll virus, PRSV—Papaya ringspot virus, PVX—Potato X virus, PVY—Potato Y virus, RHBV—Rice hoja blanca virus, RSV—Rice striped virus, RTSV—Rice tungro spherical virus, RYMV—Rice yellow mottle virus, SBWMV—wheat soil-borne mosaic virus, TMV—Tobacco mosaic virus, TSWV—Tomato spotted wilt virus, WMV 2—Watermelon mosaic 2 virus, ZYMV—Zucchini yellow mosaic virus.

The main biotechnological approach used for resistance to viral infection is the pathogen-derived resistance (PDR). This involves introduction of the pathogen-derived (PD) sequences into the plant host, which depending on the time, level and form of expression would interfere with the ability of the virus to sustain an infection.

The interaction between the transgenic sequence (PD) and the machinery for the onset of viral infection can be categorized into two groups—1 those involving the synthesis of viral proteins and 2 those involving viral RNA.

TRANSGENIC APPROACH TO VIRAL RESISTANCE INVOLVING COAT PROTEIN (CP) CODING SEQUENCE

The *Arabis mosaic nepovirus* (ArMV) causes reduction in grape yield and quality and so is important in the wine industry. ArMV has its genome on two separate molecules named RNA1 and RNA2. The coat protein (CP) is coded by RNA2. The coding sequence was amplified by PCR. The primers had restriction enzyme sites, which did not occur in the coding sequence, inserted at the 5′-end. The RNA2-cDNA was used as the template and amplified. It was ligated after restriction digestion into an *E. coli* cloning vector. A CaMV promoter was present at the 5′-end of the construct, with several transcription enhancers and at the 3′-end there was a nopaline synthase-derived transcription terminator sequence (nos). The complete construct was then digested out of the *E. coli* intermediate vector and inserted into an *Agrobacterium tumefaciens* binary vector. The transgenic plant lines obtained showed significant resistance to the ArMV infection.

Coat-protein mediated resistance (CPMR) CPMR has been one of the successes of genetic engineering. Gene encoding the coat-protein of tobacco mosaic virus (TMV), an RNA virus, has been transferred to the genome of tobacco, tomato and potato plants. These plants when challenged with TMV, showed low and delayed symptoms (Table 11.3).

"RNA effects" approach   These approaches do not target protein synthesis. They include satellite RNAs, antisense RNA and ribozyme technology. The most important effect using this approach is "gene silencing"

Satellite sequences   In plants, there are small RNA molecules that are unable to multiply in host cells without the presence of specific helper virus. Though it is not needed for the replication of the infecting virus, it can affect the symptoms caused by virus. So one of the strategies involves producing transgenic plants containing the satellite sequences. In tobacco and tomato plants carrying the cucumber mosaic virus cucumovirus (CMV) satellite RNA, a reduction in disease symptoms was found, but not strong enough to protect the plant.

A combination of satellite RNA with CMV CPMR showed stronger resistance than either strategy alone.

Antisense RNA   Introduction of an antisense construct into plants gives protection against RNA and DNA viruses. Transgenic tobacco plants expressing the antisense transcript of coat-protein genes were protected from CMV infection. Transgenic tobacco plants expressing the antisense transcript of the nucleocapsid gene were resistant to the tomato spotted wilt virus. The construct in this case expresses the RNA in the negative sense and this hybridizes with the infecting viral sequence that interferes with viral replication (Figure 11.5).

Ribozymes   These are antisense catalytic RNA molecules. These RNA molecules when expressed from the construct form partial double-stranded RNA : RNA hybrid with the large RNA, and cause its cleavage thus preventing replication.

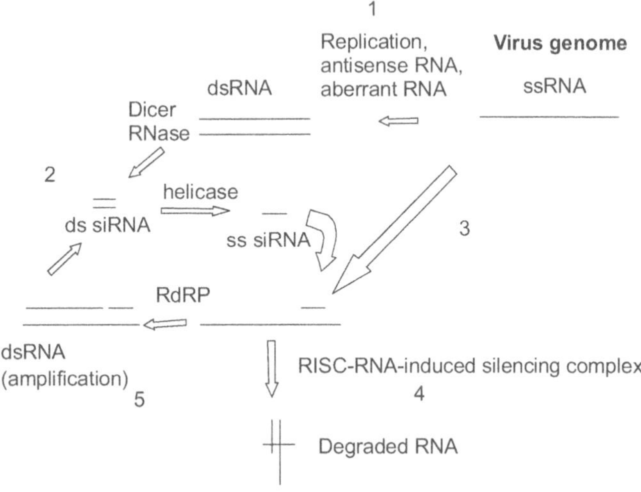

Figure 11.5 Effects of antisense RNA

STRESS–TOLERANT PLANTS

Various external factors can cause adverse effects on the growth of the plant. The stress caused by plant diseases fall under biotic stress whereas the adverse effects of environmental conditions such as water availability, temperature and salinity are classified under abiotic stress. Plants depend on internal mechanisms to tolerate the variations in the environment. Though there are different types of stresses, they result finally in three major types of damage.

1. Damage caused by water deficit
2. Damage caused by reactive oxygen species (ROS)
3. Chemical damage to cellular constituents

Resistance to Water Deficit

Water deficit stress can result due to reduction in the soil water potential and/or due to increase in the leaf water potential. This water deficit results in efflux of cellular water leading to plasmolysis and eventually cell death. It inhibits photosynthesis due to its effect on the thylakoid membranes. It also causes damage to cells because of an increase in the concentration of toxic ions.

Cells are able to counteract variation in the water potential by osmotic adjustments, i.e., by increasing the solute concentration and hence reducing the osmotic potential. The solutes produced are non-toxic compounds known as "compatible solutes", "osmolytes" or "osmoprotectants" and fall into two major categories.

1. Sugars such as trehalose and fructans, and sugar alcohols such as mannitol, sorbitol, etc.

2. Zwitterionic compounds such as amino acids, e.g. proline, and quaternary ammonium compounds, e.g. glycine betaine.

Strategies for engineering resistance to water deficit stress They focus basically on the production of osmoprotectants. This involves isolation of genes related to the biosynthetic pathways of various osmoprotectants. Constructs are then developed to express and target these compounds at appropriate location in crop plants, e.g. engineering glycine betaine production. In *Thiobacter* sp., glycine betaine is produced from choline by the action of choline oxidase (COD).

$$\text{Choline} \xrightarrow{\text{COD}} \text{Glycine betaine}$$

When this *cod* gene was transferred onto rice and *Arabidopsis* containing high levels of choline, accumulation of glycine betaine is achieved providing resistance to various water deficit stresses.

CROPS WITH IMPROVED YIELD AND QUALITY

Crop productivity can be enhanced by increasing the amount of material produced by the plant or by improving the quality of the material produced. The former is affected by photosynthetic efficiency and the harvest index or in other words dry matter allocation to the harvested part of the crop (dry matter partitioning). The latter is determined by nutritional quality, flavour, processing quality and shelf life.

Delayed Softening

An attempt was made to delay the softening of the ripe fruit so that they can be picked at a later stage. The gene coding for the wall-softening enzyme polygalacturonase was isolated and the antisense gene was synthesized. This antisense gene was transformed into plants. The antisense gene expresses RNA complementary to the polygalactouronase mRNA, which on partial hybridization with the mRNA can inhibit translation, thereby preventing the production of the enzyme.

Delayed Ripening

In tomato, ethylene biosynthesis leads to ripening. The genes of ACC synthase and ACC oxidase are involved in the ethylene biosynthetic pathway. Tomato plants transformed with the antisense genes for either of the two enzymes of ethylene biosynthesis (ACC synthase and ACC oxidase) produce no ethylene and their fruits do not ripen, but they do progress to the mature green stage. These fruits are ripened when exposed to ethylene.

Rice with Increased Vitamin A Content—Golden Rice

In tropical areas the diet is mostly insufficient in vitamin A leading to blindness in children. Golden rice is the rice that is genetically engineered to contain the needed series of genes for the biosynthesis of carotene, a precursor of vitamin A.

Manipulation of Photosynthetic Carbon Metabolism

Crop yield is dependent on the photosynthetic efficiency of the plants. In order to understand the methods involved in the manipulations of photosynthetic carbon metabolism, biosynthesis of starch has to be studied.

Pathway of starch biosynthesis The conversion of glucose 1-phosphate to ADP glucose by AGP is considered the first specific, or committed, step in starch biosynthesis (Figure 11.6).

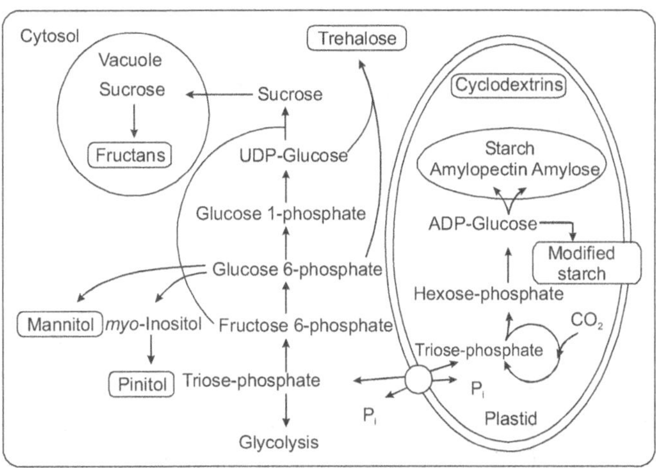

Figure 11.6 Schematic diagram of pathway for starch biosynthesis in storage organs

The key enzymes of starch biosynthesis are (1) sucrose synthase (SucSyn), (2) UDP glucose pyrophosphorylase (UGP), (3) ADP glucose pyrophosphorylase (AGP), (4) phosphoglucomutase (PGM), (5) granule-bound starch synthase (GBSS), (6) soluble starch synthase (SS), (7) starch branching enzyme (SBE), and (8) debranching enzyme (DBE). Not all alternative shunts in the pathway are shown. Fructose can be converted to glucose 1-phosphate via fructokinase (FK), phosphoglucoisomerase (PGI), and phosphoglucomutase (PGM). The relative proportions of ADP glucose synthesized in the cytoplasm and amyloplast vary from species to species.

Alteration of Starch Quantity

Storage organs constitute net consumers or sinks for photosynthetically produced carbon, whereas leaves are the sources. Generally, source–sink balances are regulated by sugar levels (hexoses as well as sucrose) and by stress. Accumulation of sugar in the leaves inhibits the source strength in tobacco. Expression of yeast invertase in the cell wall of tobacco cleaved the sucrose normally loaded into the phloem and blocked its export, mimicking a very weak sink. This work was repeated later in transgenic potato plants, and photosynthesis was shown to be inhibited by sugar accumulation in the leaves. Sucrose synthase was demonstrated through its removal in plants expressing sucrose synthase antisense under the strong 35S CaMV promoter, to play a crucial role in determining the sink strength of a potato tuber. Similar results were obtained by inhibiting the next step on the starch biosynthetic pathway, glucose 1-phosphate synthesis, through the expression of pyrophosphatase and concomitant reduction in pyrophosphate (PPi) content (Muller-Rober *et al.*, 1992).

Following the pathway further, the accumulation of both starch and protein was inhibited by antisense knockdown of AGPase levels in potato tubers. Instead, the tubers accumulated up to 30% of their dry weight as sucrose and 8% as glucose, resulting in their increased fresh weight but decreased dry weight as well as pleiotropic effects on the transcription of other genes on the starch synthetic pathway. Generally, the practical goal is to increase the accumulation of starch in tubers or grains rather than to block it. Low-starch, high-sugar potatoes would be quite poor for a major market sector, chips, crisps, and fries, because sugar accumulation results in discoloration of chips or slices during frying. The post-harvest accumulation of sugar in tubers has been limited by transgenic inhibition of UDP glucose pyrophosphorylase activity (Borovkov *et al.*, 1996). An alternative, more effective approach was taken more recently by the expression of a tobacco invertase inhibitor in tubers (Greiner *et al.*, 1999). This reduced conversion of starch to soluble sugars by up to 75%, which appears to be at levels sufficient for the practical improvement of potato processing.

Table 11.4   Recent and future developments in GM crops

Substance/ trait	Gene	Benefit	Plant
Health benefits			
Provitamin A	Phytoene synthase; phytoene desaturase; lycopene cyclase	Vitamin A supplement	Rice
Provitamin A	Phytoene desaturase	Vitamin A supplement	Tomato
Iron	Ferritin	Iron supplement	Rice
Vitamin E	γ-Tocopherol methyltransferase	Vitamin E supplement	Canola
Flavanols	Chalcone isomerase	Antioxidants, reduce risk of heart disease and cancer	Tomato
Fructans	Sucrose: sucrose fructosyl	Low-calorie alternatives to sucrose	Sugar beet

(Contd.)

Table 11.4 (Continued)

Substance/trait	Gene	Benefit	Plant
Health benefits			
cis-Stearates	Modified acyl–acyl carrier protein thioesterase	Lower risk of heart disease than *trans*-isomers	Canola
Isoflavones	Isoflavone synthase	Reduces osteoporosis, reduces blood cholesterol	*Arabidopsis*
Methionine	Antisense threonine synthase	Increased methionine levels	Potato
Seed albumin	Seed albumin gene *AmA1*	AmA1 protein is non-allergenic and rich in all essential amino acids. Increased nutritional value	Potato
Lycopene	S-Adenosyl-methionine decarboxylase	Increased lycopene levels in ripe fruit	Tomato
Yield			
Photosynthesis	Phosphoenol-pyruvate carboxylase	Improved photosynthetic efficiency, better yield	Rice

(Contd.)

Table 11.4 (Continued)

Substance/ trait	Gene	Benefit	Plant
Photosynthesis	Fructose 1,2-sedoheptulose 1,7-bisphosphate	Improved photosynthetic efficiency, better yield	Tobacco
Shade avoidance	Phytochrome A	Inhibits shade response. Plants can be planted at higher densities giving an increased yield per area	Tobacco
Photosynthesis and lifespan	Phytochrome B	Higher photosynthetic performance and longer lifespan	Potato
Dwarfing	Gibberellic acid oxidase	Inhibits GA accumulation and stem growth	Lettuce
Senescence	*ipt*	Senescence-specific expression of *ipt*, delays senescence, prolonging photosynthesis and increasing yield	Tobacco

The challenge for plant biotechnology is to convince public opinion that GM crops offer real advantages, that cannot be realistically achieved by any other way, and that the technology is safe. Only then will the world market be fully open to the GM crops and their products and the full potential realized.

Transgenic plants find a major application in the harvest of certain plant-made pharmaceuticals and technical proteins lately referred to as molecular pharming. Since this topic is vast in itself, it is discussed as a separate chapter.

REVIEW QUESTIONS

1. How is herbicide resistance engineered in plants? Explain with an example.

2. How are pest-resistant plants generated?

3. How can we engineer viral resistance in plants.

4. Write an essay on transgenic plants with altered carbon metabolism.

MOLECULAR PHARMING

INTRODUCTION

Plants have emerged as a convenient, safe and economical alternative to other expression systems over the last few years. The term **Molecular Pharming** (Molecular Farming) refers to the harnessing power of agriculture for the production of plant-made pharmaceuticals and therapeutic proteins. There is a growing demand for these protein and the available facilities are struggling to meet this demand. Molecular Pharming has the potential to provide virtually unlimited quantities of recombinant antibodies, vaccines, blood substitutes, growth factors, cytokines, and chemokines and enzymes for use as diagnostic and therapeutic tools. Plants are now gaining widespread acceptance as a general platform for the large-scale production of recombinant proteins.

There are two types of molecular farming, namely, medical and non-medical farming.

Medical molecular farming involves the production of recombinant pharmaceutical proteins, in particular antibodies. The synthesis of a pharmaceutically important protein, human growth hormone, was first described in transgenic tobacco plants in 1986.

Non-medical molecular farming includes the production of industrial products such as industrial enzymes such as laccase in transgenic maize, technical proteins for research purposes such as avidin, which is also produced in maize, milk proteins such as human beta casein, which is produced in transgenic tomatoes, and protein polymers such as collagens, which are used for medical as well as industrial purposes.

Table 12.1 lists some of the products obtained through medical and non-medical farming methods.

Table 12.1 Products of molecular farming methods

Industrial products (non-medical)	Pharmaceuticals (medical)
Proteins	Recombinant human proteins
Enzymes	Therapeutic proteins
Modified starch	Antibodies (plantibodies)
Fats	Vaccines
Oils	
Waxes	
Plastics	

EXPRESSION OF THE PROTEIN PRODUCT IN PLANTS

The expression strategies may aim for any of the three modes namely

1. Stable transformation

2. Transient transformation

3. Chloroplast transformation

In general, expression of the polypeptide depends on the stable integration of the encoding gene into the plant genome which is brought about by methods such as *Agrobacterium*-mediated transformation or particle bombardment. An advantage of this approach is that identical offsprings can be obtained by self-fertilization of the transgenic crops and consequently stable inheritance of the trait. An alternative way is by transient expression in plants via genetically engineered viruses. The number of proteins expressed using this molecular farming approach is steadily increasing, and is expected to grow rapidly as the technology matures.

The location of the expression depends on two factors, namely, the quantity and the storage time of the proteins. For instance, whole-plant expression was aimed at in tobacco with an aim of producing large amounts of protein but the preservation was poor. The seed of soy (corn) is an attractive location for long-term storage as well as for obtaining reasonable quantities of the protein.

ADVANTAGES AND DISADVANTAGES OF USING PLANTS AS BIOREACTORS

Recombinant protein production using transgenic plants as bioreactors is likely to be more economical than other systems, especially for large-scale needs. The factors in favour of plant systems as sources of animal-derived proteins, compared with other conventional methods, include:

1. the potential for the large-scale, low-cost biomass production using agriculture.

2. the low risk of contamination of the product by mammalian viruses, blood-borne pathogens, oncogenes and bacterial toxins.

3. the capacity of plant cells to correctly fold and assemble not only the antibody fragments and single-chain peptides but also the full-length multimeric proteins.

4. the ability to target proteins into intracellular compartments in which they are more stable, or even to express them directly in certain compartments (chloroplasts).

5. low downstream processing requirements for proteins administered orally.

6. elimination of the purification requirement when the plant containing the recombinant proteins is edible, such as potatoes.

7. the ability to manipulate new or multiple transgenes by sexual crossing of plants.

8. circumvention of ethical problems associated with transgenic animals.

Formulated in seeds, plant-made enzymes have been found to be an extremely convenient method for reducing storage and shipping costs, for an indefinite period of time under ambient conditions.

MEDICAL PHARMING

Pharmaceuticals

Developing drugs to treat and cure human diseases is the present-day challenge to the drug and clinical fraternity. Protein-based drugs are the fastest growing class of new drugs for treatment and prevention. But this concept of treatment has some important barriers such as capacity, cost, safety and efficacy. This blockade led to the idea of using plants as **biomachineries**.

The first pharmaceutically relevant protein made in plants was the human growth hormone, which was expressed in transgenic tobacco in 1986. In this study, the hormone was expressed as a fusion with the *Agrobacterium* nopaline synthase enzyme.

Since then, many other human proteins have been produced in an increasingly diverse range of crops. In 1989, the first antibody was expressed in tobacco (Hiatt *et al.*, 1989), which showed that plants could assemble complex functional glycoproteins with several subunits. The structural authenticity of the plant-derived recombinant proteins was confirmed in 1992, when plants were used for the first time to produce an experimental vaccine—the hepatitis B virus (HBV) surface antigen (Mason, 1992). In a further report, it is proved that the vaccine produced in tobacco plants induced the expected immune response, after it had been injected into mice (Thanavala *et al.*, 1995). Days are not too far when transgenic plants expressing vaccines would cover millions of hectares competing with agricultural crops.

Plantibodies

The term "plantibody" is derived from the key words plant and antibody, and was created to describe the products of plants that have been genetically engineered to express antibodies and antibody fragments. Antibodies can be produced *"in planta"* effectively using plants as protein factories. The term **plantibodies** was first cited in 1989. It denotes a full-length antibody molecule or a fragment of anitbody assembled within a plant.

Antibodies are glycoproteins produced in response to an infection. They can recognize a particular epitope on an antigen and also can facilitate the clearance of that antigen. There are both membrane-bound antibodies and secreted antibodies. Membrane-bound antibodies (which confer antigenic specificity on B cells) and secreted antibodies circulate in blood and are thus capable of labelling and disabling foreign entities in the body. There are five major classes of antibodies. They are IgG, IgM, IgD, IgE and IgA. The experiments of Edelman and Porter helped in elucidating the basic structure of antibody (Figure 12.1).

These antibodies have a common structure of four peptide chains. This structure consists of two identical light chains with a molecular weight of 25,000 kDa and two identical heavy chains having a molecular weight of 50,000 kDa. Each light chain is linked to the heavy chain with the help of disulphide bridges. Approximately, the first 110 amino acid residues vary greatly among antibodies of different specificity. These segments of highly variable sequences are called the **V (variable) regions**. V_L and V_H are the variable sequences present on the light chain and heavy chains respectively. All of the differences in specificity displayed by different antibodies can be traced to differences in the amino acid sequence of the V regions. And the regions of relatively constant sequence beyond the variable region have been designated as the C (constant) region (Figure 12.1).

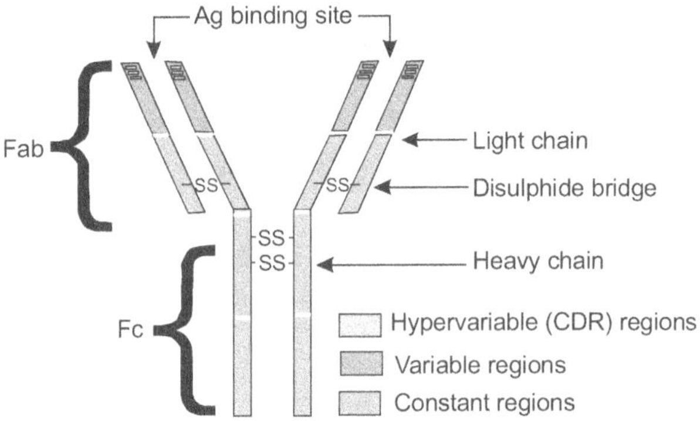

Figure 12.1 Structure of an antibody

The comparison of the amino acid sequences of the variable regions of immunoglobulins shows that most of the variability resides in three regions called the **hypervariable** regions or the **complementarity-determining** regions. Antibodies with different specificities (i.e., different combining sites) have different complementarity-determining regions while antibodies of the same specificity have identical complementarity-determining regions (i.e., CDR is the antibody-combining site). Complementarity-determining regions are found in both the H and L chains.

As the variable region of the light and heavy chains is utterly responsible for antigen binding, this region can alone be produced in plant and subsequently it can be used therapeutically without the loss of affinity. These little fragments are called single-chain variable fragments (Figure 12.2). They are rather produced by a simple fusion technology, fusing the genes for variable regions of heavy and light chains.

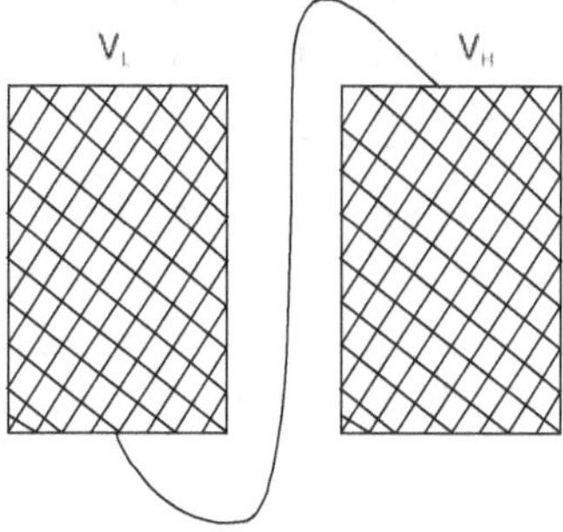

Figure 12.2   Single-chain variable fragment

Monoclonal antibodies are those that are raised in response to one specific antigen. These are desirable for human therapeutic purposes and are produced by expensive fermentation methods. An increasingly valuable option is the production of highly valuable monoclonal antibodies in plants.

Steps to produce plantibodies   Functional antibodies produced in tobacco plants were first reported over a decade and a half ago. The basic protocol that is used to generate these plantibodies involves the following steps.

1. Identifying the coding sequences of heavy and light chains on the DNA

2. Isolating the clonable sequence of heavy and light chain

3. Independent cloning of the genes of heavy and light chain in vectors (e.g. *Agrobacterium tumefaciens*)

4. Transformation of two plant cell lines (alternatively biolistic gun can also be used to introduce the foreign gene)

5. Establishment of transgenic lines, one containing the gene for heavy chain and the other the gene for the light chain

6. Crossing these two lines to generate F1 progeny. The F1 progeny would express both chains of the antibody

In a Mendelian fashion, a fully assembled and fractional antibody can be recovered from tissues in a double transgenic plant.

Production of secretory IgA   One classical example of producing functional secretory antibodies is the production of IgA. It is in fact the most abundant immunoglobulin in the body that protects the mucosal surface. IgA is a dimer that is linked together with the help of a J (joining) chain.

The process involves the production of antibody subunits in different plant lines and then they can be subsequently crossed to produce a functional antibody (Figure 12.3). These IgAs are used in preventing infection by *Streptococcus* mutants. These antibodies prevent them from colonizing by recognizing and locking the *Streptococcus* antigen I/II cell surface adhesion molecule.

The interesting feature is that in the mammalian system, both plasma and epithelial cells are involved in the production of assembled IgA, but a single transgenic plant cell is capable of producing all the different subunits efficiently.

Engineering cellular location   In mammalian cells, the heavy and light chains of antibody are produced as precursor proteins and then they are translocated into the endoplasmic reticulum (ER) under the guidance of signal peptides. Once they reach the ER, the signal peptides are cleaved, and several stress proteins such as chaperones act to bind the unassembled antibody chains and to direct subsequent folding and tetramer formation.

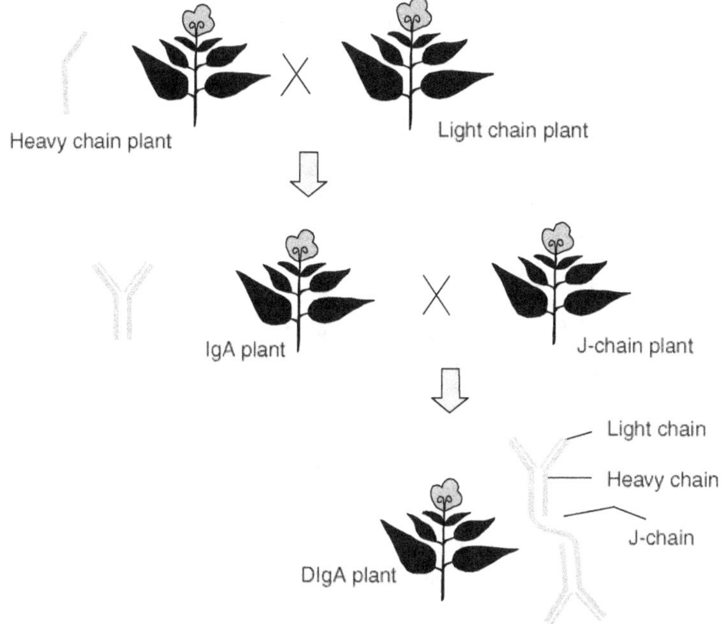

Figure 12.3   Assembly of proteins by crossing

A similar process also occurs in plant cells, and expression can be directed via signal sequences. They can be engineered to get into ER as they favour correct folding and disulphide formation. They can also be made to get secreted into apoplasm (a safe aqueous environment) or into specific plant tissues, including tubers, fruits or seeds. The engineered targeting helps in achieving higher and stable protein levels.

Glycosylation   Glycosylation is a process of adding sugar side groups to a polypeptide chain. Glycosylation in plants is found to differ from that found in mammalian cells. It is identified that plant glycans are smaller and have different terminal sugar residues.

It is speculated that the difference in glycosylation could cause lower affinity towards antigen binding and it may possibly cause the antibody to be immunogenic.

Comparison of plant and mammalian glycan biosynthesis indicates that β-1,4-galactosyltransferase is the most important enzyme missing for conversion of typical plant N-glycans to mammalian N-glycans. But as always, engineering methods answer many problems, the crossing of a plant expressing human β-1,4 -galactosyltransferase with a plant expressing heavy and light chains results in the expression of plantibodies that exhibit partially galactosylated N-glycans (approximately 30%).

Properties of plant- and mammalian-derived antibodies The following are some of the important properties of plant- and mammalian-derived antibodies.

- Identical peptide sequence.
- Identical affinity towards antigen.
- Plant system, being more versatile, can produce different types of antibodies with more stability.
- The efficiency and clearance is apparently identical in both the systems.

Post-translation processing is mildly different in plant systems, when compared to mammalian counterparts, as they have different terminal sugar residues.

Advantages The major advantages of using plantibodies as therapeutic agents are the following.

- No risk of spreading animal diseases to humans as the antibodies are produced by plants
- Inherently stable and viable proteins
- High specificity
- Low toxicity

✗   High drug approval rate

✗   Affordability and ease of production

✗   Can be injectable, topically used, or orally applied

✗   Potential long-lasting benefits

✗   Ease of purification

Uses of plantibodies to humans   The secreted antibody was topically applied to teeth and it was found to be effective for up to approximately 4 months.

Anti-herpes antibody applied to vagina of mice gave protection against the virus (Table 12.2).

Table 12.2   Therapeutic and diagnostic plantibodies

Antibody name or type	Plant	Application and specificity
Guy's 13 (secretory IgA)	*Nicotiana tabacum*	Dental caries; streptococcal
	Alfalfa	diagnostic; anti- human IgG
ScFvT84.66 (ScFv)	Wheat	Cancer treatment; carcinoembryonic antigen
T84.66 (IgG) with KDEL signal sequence	*Nicotiana tabacum*	Cancer treatment; carcinoembryonic antigen
38C13 (ScFv)	*Nicotiana benthamiana*	B-cell lymphoma treatment
CO17-1A(IgG)	*N. benthamiana*	Colon cancer; surface antigen
Anti-HSV-2 (IgG)	Soybean	Herpes simplex virus 2

Applications

1.   Antibodies expressed in plants can protect themselves from pathogens and pests by interfering with pathogen infectivity.

2. scFvs (single-chain variable fragments) generated against artichoke mottle crinkle virus gave decreased infection and delayed symptoms.

3. Full length anti-TMV antibodies help in decreasing viral necrotic lesion.

4. Tobacco plants engineered to express scFvs against the beet necrotic yellow vein virus give partial protection against virus.

5. Plants can also be engineered to express antibodies against a nematode enzyme, e.g. cellulase in saliva, thereby interfering with the feeding site of nematode and therefore inactivating the pests.

The progressive improvement of expression vectors for plantibodies, and purification strategies, as well as increase in transformable crop species is expected to lead to almost limitless availability and inexpensive recombinant immunoglobulins free of pathogens for human and animal therapy and also for novel industrial applications.

Edible Vaccines

Traditional vaccine technology despite notable successes has its limitations. Almost all vaccines now commercially available consist of either inactivated or attenuated strains of pathogens which are almost always delivered by injection (The oral polio vaccine is an exception). In contrast, many of the current vaccine development efforts focus on subunit vaccines, and these are being considered for either mucosal or parenteral delivery.

A "subunit vaccine" refers to a pathogen-derived protein or even just an epitope that cannot cause disease but can elicit a protective immune response against the pathogen. The subunit vaccine is usually a recombinant protein produced in a production host, e.g. cultured yeast cells. These proteins are injected into humans to immunize against a particular disease.

Subunit vaccines are generally considered safer to produce as they remove the need to culture the pathogenic organism and are also safe to use.

Parenteral vaccines are not effective at mucosal surfaces and therefore are not efficacious. Oral vaccines that are targeted to mucosal surfaces are able to provide better protection and stimulate both systemic as well as mucosal immune networks and also remove the need for needles. Subunit vaccines at present are being produced in fermenters and need refrigeration for transport and delivery, both of which pose constraints for developing countries. Plant edible vaccines would provide a cost-effective production system with a safe and efficacious delivery system. In plants, purification of the product may be tiresome. Attention therefore has been paid to mainly those antigens that stimulate mucosal immune system to produce secretory IgA (SIgA) at mucosal surfaces, such as gut and respiratory epithelia. In general, a mucosal response is achieved more effectively by oral instead of parenteral delivery of the antigen. Thus, an antigen produced in the edible part of a plant can serve as a vaccine against several infectious agents which invade epithelial membranes. These include bacteria and viruses transmitted via contaminated food or water, and resulting in diseases like diarrhoea and whooping cough.

To this date, most research done with plant-based vaccines have focused on diseases caused by viral infections such as hepatitis, AIDS, herpes, rabies, foot-and-mouth disease and diarrhoea. Charles Arntzen of A & M University, Texas, is a pioneer in the production of edible vaccines. In the early 1990s, he conceived of a way to solve many of the problems that bar vaccines from reaching children in developing nations. He started focusing on food that could be genetically engineered to produce vaccines in their edible parts, which could then be eaten when inoculations were needed (Figure 12.4).

Currently two forms of hepatitis B vaccine (HBV) are available, both are injectable and costly. Charles Arntzen and his group have transformed plants with the gene encoding the hepatitis B surface antigen (HBsAg); this is the same antigen used in the commercial yeast-derived vaccine. An antigenic spherical particle was recovered from these plants which is analogous to the recombinant hepatitis surface antigen (rHBsAg) derived from yeast. Parenteral immunization of mice with the plant-derived material has demonstrated that it retains both B- and T-cell epitopes, as compared to the commercial vaccine.

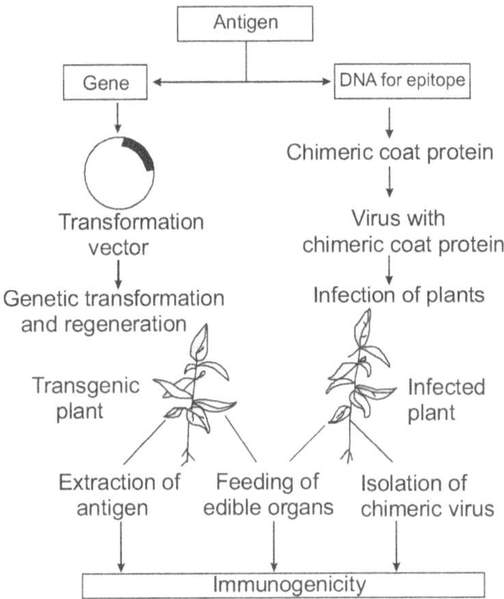

Figure 12.4   Strategies for expression of antigens in plants

Studies supported by the World Health Organization have demonstrated an effective vaccine for cholera, which provides cross-protection against enterotoxic *Escherichia coli*. This vaccine is not available, however, in large part due to cost of production of the bacterial toxin protein, which is a component of its formulation.

To address this limitation, plants were transformed with the gene encoding the β subunit of the *E. coli* heat-labile enterotoxin (LT-β) (Table 12.3). Transgenic potatoes expressing LT-β were found to induce both serum and secretory antibodies when fed to mice; these antibodies were protective in bacterial toxin assays *in vitro*. This is the first "proof of concept" for the edible vaccine.

Table 12.3 Antigens produced in transgenic plants

Protein	Plant
Hepatitis B surface antigen	Tobacco
Rabies virus glycoprotein	Tomato
Norwalk virus capsid protein	Tobacco
E. coli heat-labile enterotoxin B subunit	Potato
Cholera toxin B subunit	Potato, tobacco
Minus glutamate decarboxylase	Potato
VP1 protein of food-and-mouth disease virus	*Arabidopsis*
Insulin	Protein
Glycoprotein of swine-transmissible gastroenteritis coronavirus	*Arabidopsis*

The recombinant LT-B (rLT-β) produced in tobacco and potato showed partial pentamerization. On testing immunogenicity of rLT-β by feeding potato tubers to mice, both humoral and mucosal immune responses were reported to be stimulated. This vaccine has gone through pre-clinical trials in humans.

One of the alternative strategies of producing a plant-based vaccine is to infect the plants with recombinant viruses carrying the desired antigen that is fused to viral coat protein . The infected plants have been reported to produce the desired fusion protein in large amounts in a short time (Table 12.4). Modelska *et al.* have shown that mice injected or orally administered with the recombinant alfalfa mosaic virus (AlMV) carrying rabies peptide CPDrg 24 induced local as well as systemic immune response. Oral administration could stimulate both serum IgG as well as IgA synthesis. After immunization, 40% of the mice were protected against the challenge with a lethal dose of the virus.

Table 12.4 Transient production of antigens in plants after infection with plant viruses expressing a recombinant gene

Protein	Plant	Carrier
Influenza antigen	Tobacco	TMV
Murine zona pellucida antigen	Tobacco	TMV
Rabies antigen	Spinach	AlMV
HIV-1 antigen	Tobacco	AlMV
Mink enteritis virus antigen	Black-eyed bean	CPMV
Colon cancer antigen	Tobacco	TMV

AlMV—Alfalfa mosaic virus; TMV—Tobacco mosaic virus; CPMV —Cowpea mosaic virus.

Likewise, a 13-amino acid epitope of zona pellucida, ZP3 protein and another epitope from malarial sporozoites have been expressed as fusion proteins with TMV capsid protein with the idea of developing anti-fertility and anti-malarial vaccines. The idea of using edible vaccines against autoimmune diseases such as type I diabetes, multiple sclerosis and rheumatoid arthritis is also being closely studied.

The only demerit of transgenic plants producing vaccines is their longevity. To increase the lifespan, a regulatory promoter can be used to obtain the product upon induction, or production can be targeted exclusively to the edible regions.

Extensive research currently carried out on edible vaccines worldwide would soon overcome the obstacles and make edible vaccines an invaluable but affordable option for human beings.

NON–MEDICAL PHARMING

Industrial Enzymes

Enzymes are **bio-catalysts**, i.e., substances which accelerate a certain chemical reaction without actually being consumed. They are found in every cell of all living beings, from simple unicellular organisms to complex multicellular organisms.

Enzymes form an important part of our daily life. They are needed for many purposes such as assisting in the digestion and the catalysis of many substrates required for the well-being of our system. The strategies that have been followed for expressing enzymes in plants can be summarized as follows: 1) expression of enzymes as a truncated version in order to be small and not to provide too many sites for proteolysis 2) elimination of the expression of highly glycosylated enzymes so as to avoid glycosylation that might lead to inactivation of the protein and 3) targeting of the enzyme to the apoplastic space as it is a cellular compartment where it is expected not to be easily degraded.

For easy purification, enzymes with features different from the majority of plant proteins were used. This included, for example, thermostability and stability in acid environments.

Cellulase   Cellulose, a long chain polysaccharide made up of repeating units of α-D-glucose, is the major component of woody plants and natural fibres and considered one of the most abundant organic materials. Besides the traditional use of cellulose as raw material in the manufacturing of products such as paper and textile, it can also be used for ethanol production.

Jin *et al.* (2000) introduced the gene (*e1*) encoding the catalytic domain of thermotolerant cellulose endo 1,4-α-D-glucanase (E1) from *Acidothermus cellulolyticus* fused to a transit sequence and the 35S CaMV promoter and targeted it to the chloroplast.

The transgenic tobacco plants showed high levels of the enzyme activity proving to be a promising option for the production of this enzyme.

α-*amylase*   Amylases are a family of enzymes that hydrolyse the bond within starch molecules to produce glucose. α-amylases hydrolyse 1,4-α-glycosidic linkages in the amylose and amylopectin molecules of starch and are extensively used in a variety of industries including starch, food, brewing, alcohol, sugar, textile, and paper industries. Pen *et al.* (1992) expressed a thermostable α-amylase, isolated from *Bacillus licheniformis* in transgenic tobacco plants. A fusion containing the *B. licheniformis* α-amylase gene, *amyl* and the signal peptide of tobacco PR-S protein, which targets protein to the apoplast was transformed into tobacco protoplasts. The transgenic plants showed expression levels up to 0.3% of the total proteins. Though the expression levels were too low for commercial application, future studies are believed to improve the yield.

Bioplastics and Biotechnology

The growing mountain of plastic waste is a major environmental concern of the world and scientists are focusing on production of plastic that is biodegradable to address this problem. The other aspect is that conventional plastics are produced from oil which causes further reduction on petroleum stores that is running low.

Polyhydroxybutyrate   Polyhydroxyalkanoates (PHAs) are a class of linear polyesters naturally produced by bacteria through fermentation of sugars or lipids. The most common type of PHA is polyhydroxybutyrate (PHB), that can act as biodegradable plastic. Attempts are being made by researchers to produce these in plants such as *Arabidopsis*, mustard, corn, potato, sugar cane and tobacco though bacteria were earlier used. In bacteria the production cost was high compared to synthetic plastics. PHB is biosynthesized by bacteria using the substrate acetyl-CoA through the action of three enzymes encoded by *phb*A (α-ketothiolase), *phb*B (acetoacetyl-CoA reductase), and *phb*C (PHB synthase). In transgenic rice and tobacco plants, these genes have been successfully expressed.

Through *Agrobacterium*-mediated transformation, the transgenic tobacco plants showed PHB accumulation up to 3.2 mg/g of dry weight. In another research aiming at the production of PHB in transgenic tobacco, the polycistronic *phb* operon was transferred to tobacco plastome using biolistic transformation methods. The researchers observed that the highest PHB accumulation (1.7% dry weight) was achieved only during the early stages of *in vitro* culture. PHB production in transplastomic tobacco plants decreased over time and mature plants showed stunted growth and male sterility. While these expression levels are still not ideal for large-scale commercial production of PHB, these experiments laid the foundation for achieving the goal of producing bioplastics in plants.

Polyhydroxy butyrate valerate PHBV Poly (3-hydroxybutyrate-co-3-hydroxyvalerate) or PHBV is produced naturally by some species of bacteria and it is found that it can be heat-formed into flexible plastic suitable for many applications where biodegradable plastics are desirable. Genetic engineering techniques are being employed to increase the production efficiency of these compounds. There are five enzymes, two from the plant and three from bacteria, which when combined, produce PHBV. Of the two plant enzymes, one is currently produced in the mitochondria. Research is going on to modify the plant so that this enzyme is diverted to the chloroplasts. Once this is achieved, bacterial enzymes can be introduced to produce PHBV in chloroplasts. Scientists used genetic engineering techniques to insert these genes into the DNA of a number of plants belonging to the mustard family.

REVIEW QUESTIONS

1. What are plantibodies and how are they generated?

2. How are plantibodies useful to humans?

3. Write a note on plant as bioreactor.

4. Write an essay on molecular pharming.

5. What are edible vaccines?

GLOSSARY

ABA-response element (ABRE) A six-nucleotide sequence element found in the promoters of genes regulated by abscisic acid (ABA), a plant hormone

Abscissic acid (ABA) A plant harmone also known as abscisin II and dormin, is a plant hormone. It functions in many plant developmental processes, including bud dormancy.

Abscission The shedding of leaves, flowers and fruits from a living part. The process where by specific cells in the leaf petiole differentiate to form an abscission layer, allowing a dying/dead organ to separate from the plant.

ACC oxidase An enzyme involved in ethylene biosynthesis. It catalyses the conversion of ACC (amino cyclopropane carboxylic acid) to ethylene.

ACC synthase An enzyme that catalyses the synthesis of ACC from S-adenosylmethionine.

Activator A DNA-binding protein that regulates one or more genes by increasing the rate of transcription.

Adenosine-5'-triphosphate A multifunctional nucleotide. The structure of this molecule consists of a purine base (adenine) attached to the 1' carbon atom of a pentose sugar (ribose). Three phosphate groups are attached at the 5' carbon atom of the pentose sugar.

Agonist A type of ligand or drug that binds and alters the activity of a receptor. The ability to alter the activity of a receptor, also known as the agonist's efficacy is a property that distinguishes it from antagonists, a type of receptor ligand which also bind a receptor but which do not alter the activity of the receptor.

Amino acid an amino acid is a molecule containing both amine and carboxyl functional groups.

Aminoacyl-tRNA synthetase An enzyme that catalyzes the esterification of a specific amino acid or its precursor to one of all its compatible cognate tRNAs to form an aminoacyl-tRNA.

Antenna complex A group of pigment molecules that cooperate to absorb light energy and transfer it to a reaction centre complex.

Antenna pigments Pigments found in the antenna complex such as chlorophylls and carotenoids.

Anthocyanin Water-soluble vacuolar pigments that may appear red, purple, or blue according to pH.

Antibodies Gamma globulin proteins that are found in blood or other bodily fluids of vertebrates, and are used by the immune system to identify and neutralize foreign objects, such as bacteria and viruses.

Anticodon A unit made up of three nucleotides that correspond to the three bases of the codon on the mRNA. Each tRNA contains a specific anticodon triplet sequence that can base-pair to one or more codons for an amino acid.

Antigen (immunogen) A substance that prompts the generation of antibodies and can cause an immune response. The word originated from the notion that they can stimulate antibody generation.

Antisense RNA Single-stranded RNA that is complementary to a messenger RNA (mRNA) strand transcribed within a cell. Antisense RNA may be introduced into a cell to inhibit translation of a omplementary mRNA by base pairing to it and physically obstructing the translation machinery.

Apical dominance Phenomenon whereby the main central stem of the plant is dominant over (i.e., grows more strongly than) other side stems.

Apoptosis Programmed cell death showing characteristic morphological and biochemical changes, including fragmentation of nuclear DNA between the nucleosomes. It occurs in some senescing plant tissues, differentiatiing xylem tracheary elements and in the hypersensitive response against pathogens.

Arabidopsis A plant belonging to the family Brassicaceae. It is a small flowering plant related to cabbage and mustard.

ATP synthase (ATPase or CF_0-CF_1) An enzyme that synthesizes ATP from ADP and phosphate (P). It consists of two parts: a hydrophobic membrane-bound portion (CF_0) and a portion that sticks out into the stroma (CF_1).

Auxin A plant growth substance that regulates stem growth. It induces cell division in callus tissue in the presence of cytokinin, lateral root formation at the cut surfaces of stems, and parthenocarpic growth and ethylene formation in fruits.

Biolistic gun Also known as gene gun or the biolistic particle delivery system and originally designed for plant transformation, this is a device for injecting cells with genetic information. The payload is an elemental particle of a heavy metal coated with plasmid DNA.

Bioplastics A form of plastics derived from renewable biomass sources, such as vegetable oil, corn starch, pea starch or microbiota, rather than traditional plastics which are derived from petroleum. They are used either as a direct replacement for traditional plastics or as blends with traditional plastics.

C_3 plants Plants in whcich the first stable product of CO_2 fixation is a three-carbon compound (i.e., 3-phosphoglycerate).

C_4 plants Plants in which the first stable product of CO2 assimilation in mesophyll cells is a four-carbon compound that is immediately transported to bundle sheath cells and decarboxylated.

Calvin cycle (Calvin-Benson-Bassham cycle or carbon fixation) A series of biochemical reactions that takes place in the stroma of chloroplasts in photosynthetic organisms. It is one of the light-independent reactions or dark reactions.

Cambium A layer or layers of tissue, also known as lateral meristems, that are the source of cells for secondary growth. There are two types of cambiumcork cambium and vascular cambium.

Capping A process of adding 7-methylguanosine to the 5′-end of eukaryotic mRNA.

Central dogma Dogma of molecular biology that deals with the detailed residue-by-residue transfer of sequential information. It states that information cannot be transferred back from protein to either protein or nucleic acid.

Chlorophyll A group of light-absorbing green pigments active in photosynthesis and found in most plants, algae, and cyanobacteria. It absorbs light, most strongly in the blue and red but poorly in the green portions of the electromagnetic spectrum, hence the green colour of chlorophyll-containing tissues like plant leaves.

Chloroplasts Organelles found in plant cells and eukaryotic algae that conduct photosynthesis.

Chromatin The complex of DNA and protein that makes up chromosomes. It is found inside the nuclei of eukaryotic cells, and within the nucleoid in prokaryotic cells.

Citric acid cycle A series of enzyme-catalysed chemical reactions of central importance in all living cells that use oxygen as part of cellular respiration. It also known as the tricarboxylic acid cycle (TCA cycle) or the Krebs cycle.

Codon A triplet nucleotide sequence of an mRNA that codes the amino acid of a protein.

Coleoptile The pointed protective sheath covering the emerging shoot in monocotyledons such as oats and grasses.

Crassulacean acid metabolism (CAM) An elaborate carbon fixation pathway in some plants. These plants fix carbon dioxide (CO_2) during the night, storing it as the four carbon acid malate.

Cristae Folds in the inner mitochondrial membrane that project into the mitochondrial matrix. The enzymes of the electron transport chain and of oxidative phosphorylation are localized in the cristae.

Critical day length The minimum length of the day required for flowering of a long-day plant. The maximum length of day that will allow short-day plants to flower. However, studies have shown that it is the length of the night, not the length of the day, this is important.

Critical night length The night length must be exceeded for flowering of short-day plants, or for inhibition of flowering in long-day plants.

Cytochromes c A peripheral, mobile component of the mitochondrial electron transport chain that oxidizes complex III and reduces complex IV.

Cytochromes Membrane-bound hemoproteins that contain heme groups and carry out electron transport.

Cytokinins (CK) A class of plant growth substances (plant hormones) that promote cell division. They are primarily involved in cell growth, differentiation and other physiological processes.

Cytoplasm Contents of a cell enclosed within the plasma membrane. In eukaryotic cells the cytoplasm contains organelles, such as mitochondria that are filled with liquid and kept separated from the cytoplasm by cell membranes.

Cytoplasmic male sterility (CMS) A maternally inherited male-sterile phenotype caused by mutation in mtDNA, in which viable pollen is not formed.

Cytosol Colloidal aqueous phase of the cytoplasm containing dissolved solutes but excluding supramolecular structures such as ribosomes and components of the cytoskeleton.

Dark reactions Chemical reactions that convert carbon dioxide and other compounds into glucose. It occurs in the stroma, the fluid filled area of a chloroplast outside of the thylakoid membranes. These reactions, unlike the light-dependent reactions, do not need light to occur; hence the term *dark reactions*.

Deoxyribonucleic acid (DNA) A nucleic acid that contains the genetic instructions used in the development and functioning of all known living organisms and some viruses.

Diterpenes Terpenes having 20 carbons, four five-carbon isoprene units.

DNA polymerase An enzyme that assists in DNA replication that catalyze the polymerization of deoxyribonucleotides along the side of a DNA strand, which they "read" and use as a template.

Dormancy A period in an organism's life cycle when growth, development, and (in animals) physical activity is temporarily suspended. This minimizes metabolic activity and therefore helps an organism to conserve energy. Dormancy tends to be closely associated with environmental conditions.

Ecdysone A steroidal pro-hormone of the major insect molting hormone 20-hydroxyecdysone, which is secreted from the prothoracic glands. Moulting hormones (ecdysone and its homologues) in insect are generally called ecdysteroids.

Electron transport chain (in the mitochondrion) A series of protein complexes in the inner mitochondrial membrane linked by the mobile electron carriers ubiquinone and cytochrome c, which catalyse the transfer of electrons from NADH to O_2. In the process a large amount of free energy is released. Some of that energy is conserved as an electrochemical proton gradient.

Elongation factors A set of proteins that facilitate the events of translational elongation, the steps in protein synthesis from the formation of the first peptide bond to the formation of the last one.

Embryo Part of a seed consisting of precursor tissues for the leaves, stem and root, as well as one or more cotyledons.

Endoplasmic reticulum (ER) An organelle found in all eukaryotic cells that is an interconnected network of tubules, vesicles and cisternae.

Endosperm Tissue produced in the seeds of most flowering plants around the time of fertilization. It surrounds the embryo and provides nutrition in the form of starch, though it can also contain oils and protein. This makes endosperm an important source of nutrition in human diet.

Enhancer A short region of DNA that can be bound with proteins (namely, the *trans*-acting factors, much like a set of transcription factors) to enhance transcription levels of genes (hence the name) in a gene cluster.

Ethylene A gaseous hydrocarbon hormone that is synthesized from the amino acid methionine via ACC. It regulates fruit ripening and plays a variety of roles such as in senescence of leaves and fruit, elongation of roots, and responses to waterlogging and other stresses.

Etiolated seedlings Dark-grown seedlings in which the hypocotyl and stem are more elongated, cotyledons and leaves do not expand, and chloroplasts do not mature.

Etiolation Growth of seedlings in darkness and in which a seedling or organ rapidly elongates without production of chloroplast.

Eukaryotes Animals, plants, fungi and protists whose cells are organized into complex structures enclosed within membranes. Many eukaryotic cells contain other membrane-bound organelles such as mitochondria, chloroplasts and Golgi bodies.

Exon A sequence that is translated into proteins.

Expression vector A plasmid-expression construct. It is used to introduce and express a specific gene into a target cell. Expression vector allows production of large amounts of stable mRNA. Once the expression vector is inside the cell, the protein that is encoded by the gene is produced by the cellular transcription and translation machinery.

Ferredoxins A small, water-soluble, iron-sulphur protein that mediates electron transfer in a range of metabolic reactions.

Florigen / flowering hormone The terms used for the hypothesized hormonelike molecules that control and/or trigger flowering in plants.

Genetically modified (GM) foods Food products that have

had their DNA directly altered through genetic engineering.

Genotype Genetic constitution of a cell, an organism, or an individual.

Gibberellin A₁ (GA₁) A chemically distinct form of gibberellin. The primary GA1, is active in stem growth for most species.

Gibberellin A₃ (GA₃) The main gibberellin found in fungal cultures; commonly available and used in the management of food crops, the malting of barley, and the extension of sugar cane (to increase sugar yield). It occurs rarely in plants.

Gibberellin (GAs) Plant hormones that regulate growth and influence various developmental processes, including stem elongation, germination, dormancy, flowering, sex expression, enzyme induction and leaf and fruit senescence.

Glucocorticoids (GC) A class of steroid hormones characterized by an ability to bind with the glucocorticoid receptor (GR) and trigger similar effects. These may be either slow, mediated genomically through nuclear receptors, or fast, mediated nongenomically through membrane-associated receptors and signaling cascades.

Gluconeogenesis Synthesis of carbohydrates through the reversal of glycolysis.

Glucuronidases Enzymes that separate glucuronic acid molecules from other molecules by cutting glycosidic bonds. They are thus classified as glycoside hydrolases that cleave glucuronides.

Glycolysis Sequence of reactions that convert glucose into pyruvate with the concomitant production of a relatively small amount of adenosine triphosphate (ATP).

Glycosylation Enzymatic process that links saccharides to produce glycans, either free or attached to proteins and lipids. This enzymatic process produces one of four fundamental components of all cells (along with nucleic acids, proteins, and lipids).

Golgi complex Also called the golgi body, Golgi apparatus, or dictyosome, this is an organelle found in most eukaryotic cells that is composed of membrane-bound stacks known as cisternae.

Grana (Singular granum) A stack of thylakoid discs that are connected by stroma thylakoids, also called intergrana thylakoids or lamellae. Chloroplasts can have from 10 to 100 grana.

Gynoecium Female reproductive part of a flower. A gynoecium is composed of one or more pistils. A pistil may consist of a single free carpel, or be formed from a number of carpels that are fused. The pistil itself is formed from the stigma, style, and ovary.

Histones Chief protein components of chromatin.

Hormone response element (HRE) A response element for hormones, a short sequence of DNA within the promoter of a gene that is able to bind a specific hormone-receptor complex and therefore regulate transcription. The sequence is most commonly a pair of inverted repeats separated by three nucleotides, which also indicates that the receptor binds as a dimer.

Indole 3-acetic acid(IAA) A member of the group of phytohormones called auxins. IAA is generally considered to be the most important native auxin.

Inducer A molecule that stimulates gene expression.

Introns Sequences in a gene that are not translated into proteins. Also called intervening sequences.

Kinase A type of enzyme that transfers phosphate groups from high-energy donor molecules, such as ATP, to specific target molecules (substrates); the process is termed *phosphorylation*. This enzyme is alternatively known as a phosphotransferase.

Kinetin A kind of cytokinin, a class of plant hormone that promotes cell division.

Ligand A substance that is able to bind to and form a complex with a biomolecule to serve a biological purpose.

Light reaction The initial stage of the photosynthetic system is the light-dependent reaction, which converts solar energy into potential energy.

The light dependent reaction produces oxygen gas and converts ADP and NADP+ into the energy carriers ATP and NADPH.

Messenger ribonucleic acid (mRNA) A molecule of RNA that is transcribed from a DNA template, and carries coding information to the sites of protein synthesis, the ribosomes.

Micropropagation The practice of rapidly multiplying stock plant material to produce a large number of progeny plants, using modern plant tissue culture methods.

Mitochondrion (plural mitochondria) is a membrane-enclosed organelle found in most eukaryotic cells. The organelle is composed of compartments that include the outer membrane, the intermembrane space, the inner membrane, and the cristae and matrix.

Molecular farming (Officialy known as Transgenic non-food GM plant pharming and biopharming). This is a type of geneteic modification involving the use of plants, and potentially also animals, as the means to produce compounds of therapeutic value. The idea is to use such crops as biological factories to generate drugs that are difficult or expensive to produce in any other way. The issue of genetically modified crops has been around for a number of years and continues to be a controversial subject.

mtDNA Mitochondrial DNA or genome of mitochondria. Plant mitochondrial genome is much larger than that of mammals and yeasts, and contains large regions of non-coding DNA.

NADP Nicotinamide adenine dinucleotide phosphate ($NADP^+$) is used in anabolic reactions, such as lipid and nucleic acid synthesis, which require NADPH as a reducing agent. NADPH is the reduced form of $NADP^+$, and $NADP^+$ is the oxidized form of NADPH.

Naphthaleneacetic acid (NAA) An organic compound with the formula $C_{10}H_7CH_2CO_2H$. NAA is a plant hormone in the auxin family and is an ingredient in many commercial plant rooting horticultural products; it is a rooting agent and used for the vegetative propagation of plants from stem and leaf cutting. It is also used for plant tissue culture.

Non-coding RNA (ncRNA) RNA molecule that is not translated into a protein.

Nuclear receptors A class of proteins found within the interior of cells that are responsible for sensing the presence of hormones and certain other molecules. In response, these receptors work in concert with other proteins to regulate the expression of specific genes thereby controlling the development, homeostasis, and metabolism of the organism.

Nucleolus A membraneless component of the cell nucleus. The main function of the nucleolus is the biogenesis and assembly of ribosome components (rRNA, ribosomal proteins).

Nucleosomes Fundamental repeating units of eukaryotic chromatin. The nucleosome core particle consists of approximately 147 base pairs of DNA wrapped around a histone octamer consisting of 2 copies each of the core histones H2A, H2B, H3, and H4.

Nucleotides Organic compounds that consist of three joined structures: a nitrogenous base, a sugar, and a phosphate group. They are the structural units of RNA and DNA.

Nucleus The cell's-control center that contains the chromosomal DNA.

Okazaki fragment Short fragment of DNA (with an RNA primer at the 5' terminus) created on the lagging strand during DNA replication. The lengths of Okazaki fragments is 1,000 to 2,000 nucleotides in *E. coli* and are generally 100 to 200 nucleotides in eukaryotes.

Origin of replication Particular sequence in a genome at which replication is initiated. This can either be DNA replication in living organisms such as prokaryotes and eukaryotes, or RNA replication. It also called the replication orgin.

parthenocarpy The natural or artificially induced production of fruit without fertilization of ovules. The fruit is therefore seedless.

Permeases Membrane-transport proteins, a class of multipass transmembrane protein, which facilitate the diffusion of a specific molecule in or out of the cell.

Petal A highly-modified part of the corolla of a flower. Corolla is the collective name for all of the petals of a flower.

Phenotype Any observable characteristic of an organism, such as its morphology, development, biochemical or physiological properties, or behavior.

Phosphorylation Addition of a phosphate (PO_4) group to a protein molecule or a small molecule.

Photoexcitation The mechanism of electron excitation by photon absorption, when the energy of the photon is too low to cause photoionization. The absorption of photon takes place in accordance to the Planck's Quantum Theory.

Photolysis A reaction in which a chemical compound is broken down by photons. It is otherwise called as photodissociation/ photodecomposition.

Photomorphogenesis The influence and specific roles of light on plant growth and development. In the seedling, it refers to light-induced changes in gene expression to support above-ground growth in the light rather than below-ground growth in the dark.

Photoperiod The amount of time per day that a plant is exposed to light or darkness. It mat control various aspects of sexual or vegetative reproductive development including flowering and tuberization.

Photoreceptor A chromo-protein that responds to being exposed to a certain wavelength of light by initiating a signal transduction cascade.

Photorespiration Uptake of atmospheric O_2 with a concomitant release of CO_2 by illuminated leaf. Also, the alternate pathway for production of glyceraldehyde 3-phosphate (G3P) by RuBisCO, the main enzyme of the light-independent reactions of photosynthesis (also known as the Calvin cycle or the Primary carbon reduction cycle). Although RuBisCO favors carbon dioxide to oxygen, (approximately 3mcarboxylations per oxygenation), oxygenation of RuBisCO occurs frequently, producing a glycolate and a glycerate.

Photosynthate The carbon-containing products of photosynthesis.

Photosynthesis A metabolic pathway that converts light energy into chemical energy. Its initial substrates are carbon dioxide and water; the energy source is sunlight (electro-magnetic radiation); and the end-products are oxygen and (energy-containing) carbohydrates, such as sucrose, glucose or starch.

Photosynthetic electron transport Electron flow from light-excited chlorophyll and the oxidation of water, through PS II and PS I, to the final electron acceptor $NADP^+$.

Photosystem A functional unit of chloroplast that harvests light energy to provide for power electron transfer and to generate a proton motive force used to synthesize ATP.

Photosystem I (PS I) A system of photoreactions that absorbs maximally far-red light (700 nm), oxidizes plastocyanin and reduces ferredoxin.

Photosystem II (PS II) A system of photoreactions that absorbs maximally red light (680 nm), oxidizes water and reduces plastoquinone.

Phototropism One of the many plant tropisms or movements which respond to external stimuli. Growth towards a light source is a positive phototropism, while growth away from light is called negative phototropism (or Skototropism).

***Phy* genes** The genes for the apoproteins of phytochrome. They belong to phytochrome gene family. In Arabidopsis, the *phy* genes are *phy*A, *phy*B, *phy*C, *phy*D and *phy*E.

Phytoalexins Antibiotics produced by plants that are under attack. Phytoalexins tend to fall into several classes including terpenoids, glycosteroids and alkaloids.

Phytochrome A plant growth-regulating photoreceptor protein that primarily absorbs red light and far-red light, and also blue light. It is a holoprotein that contains the chromophore, phytochromobilin.

Phytohormones Substances that influence plant growth and development at low concentrations. Major classes are abscisic acid, auxin, cytokinin, ethylene and gibberellin.

Plasma membrane The cell membrane (also called the plasmalemma, or phospholipid bilayer) that is selectively permeable lipid bilayer found in all cells. It contains a wide variety of biological molecules, primarily proteins and lipids.

Plasmid   (ccc) Small, covalently closed, circular, extrachromosomal bacterial DNA that can replicate independently from the bacterial chromosome.

Plasmodesmata Microscopic channels which traverse the cell walls of plant cells and enable transport and communication between them.

Plastocyanin Copper-containing protein involved in electron transfer.

Pollen A fine to coarse powder consisting of microgametophytes (pollen grains), which produce the male gametes (sperm cells) of seed plants.

Polyadenylation Synthesis of a poly(A) tail, a stretch of RNA where all the bases are adenines, at the end of an RNA molecule.

Polyhydroxybutyrate (PHB) A polymer belonging to the polyesters class. PHB is produced by microorganisms (like *Alcaligenes eutrophus* or *Bacillus megaterium*) apparently in response to conditions of physiological stress. It is primarily a product of carbon assimilation (from glucose or starch) and is employed by microorganisms as a form of energy-storage molecule to be metabolized when other common energy sources are not available.

Prokaryotes A group of organisms that lack a cell nucleus (= karyon), or any other membrane-bound organelles.

Promoter A regulatory region of DNA generally located upstream (towards the 5'-region of the sense strand) of a gene that allows transcription of the gene.

Proteins Large organic compounds made up of amino acids arranged in a linear chain and joined together by peptide bonds between the carboxyl and amino groups of adjacent amino acid residues.

Replicon A DNA or RNA molecule, or a region of DNA or RNA, that replicates from a single origin of replication.

Repressor DNA-binding protein that regulates the expression of one or more genes by decreasing the rate of transcription. This blocking of expression is called repression.

Respiration The complete oxidation of carbon compounds to CO2 and H2O, using oxygen as the final electron acceptor. Energy is released and conserved as ATP.

Ribonucleic acid (RNA) A nucleic acid consisting of a long chain of nucleotide units. Each nucleotide consists of a nitrogenous base, a ribose sugar, and a phosphate.

Ribosomal RNA (rRNA) Central component of the ribosome, the protein manufacturing machinery of all living cells. The function of the rRNA is to provide a mechanism for decoding mRNA into amino acids and to interact with the tRNAs during translation by providing peptidyl transferase activity.

Ribozyme RNA molecule that catalyzes a chemical reaction. Many natural ribozymes catalyze either the hydrolysis of one of their own phosphodiester bonds or the hydrolysis of bonds in other RNAs. It is also called as RNA enzytme or catalytic RNA.

RNA polymerase (RNAP or RNApol) An enzyme that synthesizes RNA. In cells, RNAP is needed for constructing RNA chains from DNA genes as templates, a process called transcription.

Root cap A section of tissue at the tip of a plant root. Root caps contain statoliths which are involved in gravity perception in plants. If the cap is carefully removed the root will grow randomly. The root cap protects the growing tip in plants.

RuBisCO   Ribulose-1,5-bisphosphate carboxylase / oxygenase) An enzyme that is used in the Calvin cycle to catalyze the first major step of carbon fixation, a process by which the atoms of atmospheric carbon dioxide are made available to organisms in the form of energy-rich molecules such as sucrose. RuBisCO catalyzes either the carboxylation or oxygenation of ribulose 1,5-bisphosphate (also known as *RuBP*) with carbon dioxide or oxygen.

***Scr* gene** *Arabidopdsis SCARECROW* gene that controls tissue organization and cell differentiation in the embryo, hypocotyl, primary roots and secondary roots.

***Scr* mutant** Arabidopsis mutant in which hypocotyl and inflorescence are agravitrophic and lack both endodermis and starch sheath.

senescence An active, genetically controlled, developmental process in which cellular structures and macromolecules are broken down and translocated away from the senesing organ (typically leaf) to actively growing regions that serve as nutrient sinks. This was initiated by environmental cues, and regulated by hormones.

Sepal A part of the flower of angiosperms or flower plants. Sepals in a "typical" flower are green and lie under the more conspicuous petals. As a collective unit the sepals are called the *calyx* of a flower.

Signal transduction Any process by which a cell converts one kind of signal or stimulus into another. Most processes of signal transduction involve ordered sequences of biochemical reactions inside the cell, which are carried out by enzymes, activated by second messengers, resulting in a *signal transduction pathway*.

SnRNPs Particles that combine with pre-mRNA and various proteins to form spliceosomes (a type of large molecular complex). SnRNPs "recognize" the places along a strand of pre-mRNA and are essential in the removal of introns. These molecules are found within the cell's nucleus.

Solenoid   The packing of DNA as a 30-nm fiber of chromatin and results from the helical winding of at least five nucleosome strands.

Spliceosome   A complex of specialized RNA and protein subunits that removes introns from a transcribed pre-mRNA (hnRNA) segment. This process is generally referred to as splicing.

Splicing   A modification of an RNA after transcription, in which introns are removed and exons are joined.

Stamen   Male organ of a flower. Each stamen generally has a stalk called the filament (from Latin *filum*, meaning "thread"), and, on top of the filament, an anther and pollen sacs, called *microsporangia*.

Steroid   A steroid is a terpenoid lipid characterized by a carbon skeleton with four fused rings.

Stoma   A tiny opening or pore, found mostly on the underside of a plant leaf and used for gas exchange.

Systemin   Recently discovered plant hormone involved in wound response. It is unique from other plant hormones as it is a peptide.

TATA box   Also called Goldberg-Hogness box, this is a DNA sequence (*cis*-regulatory element) found in the promoter region of most genes in eukaryotes and Archaea.

Template strand   One strand of the DNA serving as a template for the synthesis of a complementary strand of RNA. The template DNA strand is called the transcribed strand with antisense sequence and the mRNA transcript is said to be sense sequence (the complement of antisense). Because the DNA is double-stranded, the strand complementary to the antisense sequence is called non-transcribed strand and has the same sense sequence as the mRNA transcript (though T bases in DNA are substituted by U bases in RNA).

Termination codon (stop codon)   A nucleotide triplet within messenger RNA that signals a termination of translation. They are UAA, UAG and UGA.

Terminator A section of genetic sequence that marks the end of gene or operon on genomic DNA for transcription.

Thylakoid A membrane-bound compartment inside the chloroplasts and cyanobacteria. They are the site of the light-dependent chemical reactions of photosynthesis. Thylakoid consists of a thylakoid membrane surrounding a thylakoid lumen.

Totipotency Ability of a single cell to divide and produce all the differentiated cells in an organism, including extra-embryonic tissues.

Transcription Process of copying DNA to RNA by an enzyme called RNA polymerase (RNAP).

Transcription factor A protein that binds to specific parts of DNA using DNA binding-domains and is part of the system that controls the transfer (or transcription) of genetic information from DNA to RNA. It is also known as sequence-specific DNA binding factor.

Transcription terminator *See* Terminator

Transfer RNA (tRNA) A small RNA (usually about 74–95 nucleotides) that transfers a specific amino acid to a growing polypeptide chain at the ribosomal site of protein synthesis during translation.

Transgenic Plant A plant expressing a foreign gene introduced by genetic engineering techniques.

Translation The first stage of protein biosynthesis (part of the overall process of gene expression). In translation, messenger RNA (mRNA) is decoded to produce a specific polypeptide according to the rules specified by the genetic code.

Transposase An enzyme that binds to single-stranded DNA and that can incorporate a transposon into genomic DNA.

Transposons Sequences of DNA that can move around to different positions within the genome of a single cell. The process called transposition. In the process, they can cause mutations in the genome.

Transposons are also called as "jumping genes".

Tryptophan One of the 20 standard amino acids, as well as an essential amino acid in the human diet (abbreviated as Trp or W).

Upstream Upstream and downstream both refer to a relative position in DNA or RNA. Each strand of DNA or RNA has a 5'-end and a 3'-end, so named for the carbons on the deoxyribose ring. Relative to the position on the strand, downstream is the region towards the 3'-end of the strand. Since DNA strands run in opposite directions, downstream on one strand is upstream on the other strand.

Vaccine Biological preparation which is used to establish or improve immunity to a particular disease. Vaccines can be prophylactic (e.g. to prevent or ameliorate the effects of a future infection by any natural or "wild" pathogen), or therapeutic (e.g. vaccines against cancer are also being investigated).

Zeatin A naturally occurring cytokinin that stimulates mature plant cells to divide when added to culture medium along with an auxin. Chemically known as trans-6-(4-hydroxy 3-methylbut 2-enlyamino) purine.

Zeatin ribosides Zeatin with a ribose attached to the amino purine moiety. It is the main cytokinin in the xylem exudates.

Zinc finger A large superfamily of protein domains that can bind to DNA. A zinc finger consists of two antiparallel β-strands, and an α–helix. The zinc ion is crucial for the stability of this domain type in the absence of the metal ion. The domain unfolds as it is too small to have a hydrophobic core.

Zygote A single diploid cell formed by fusion of a female gemete (ovum) and a male gamete (sperm) by the process of fertilization.

REFERENCES

Adrian Slater, Nigel Scott and Mark Fowler. (2003). *Plant Biotechnology: The Genetic Manipulation of Plants*. Oxford University Press, Oxford.

Ainley, W.M. and Key, J.L. (1990). "Development of a heat-shock inducible expression cassette for plants: characterization of parameters for its use in transient expression assays." *Plant Molecular Biology*. **14**: 949–967.

Allan, A.C., Fricker, M.D., Ward, J.L., Beale, M. and Trewavas, A.J. (1994). "Two transduction pathways mediate rapid effects of abscisic acid in *Commelina* guard cells." *Plant Cell*. **6**: 1319–1328.

Arntzen, C. (1990). "Vaccines expressed in plants." US Patent 6136320. Issued on October 24, 2000.

Baldwin, I.T., Schmelz, E.A. and Ohnmeiss, T.E. (1994). "Wound-induced changes in root and shoot jasmonic acid pools correlate with induced nicotine synthesis in *Nicotiana sylvestris* Epegazzini and Comes." *Journal of Chemical Ecology*. **20**: 2139–2157.

Baydoun, E. and Fry, S. (1985). "The immobility of pectic substances injured tomato leaves and its bearing on the identity of the wound hormone." *Planta*. **165**: 269–276.

Borovkov. A.Y., McClean, P.E., Sowokinos, J.R., Ruud, S.H. and Secord, G.A. (1996). "Effect of expression of UDP-glucose pyrophosphorylase ribozyme and antisense RNAs on the enzyme activity and carbohydrate composition of field-grown transgenic potato plants." *Journal of Plant Physiology*. **147**: 644–652.

Borthwick, H.A., Hendricks, S.B., Parker, M.W., Toole, E.H. and Toole, V.K. (1952). "A reversible photoreaction controlling seed germination." Proc. Natl. Acad.Sci., USA. **38**: 662–666.

Bradley, D.J., Kjellbom, P. and Lamb, C.J. (1992). "Elicitor and wound-induced oxidative cross-linking of a praline-rich plant cell wall protein: a novel, rapid defense response." *Cell.* **70**: 21–30.

Brand, U., Fletcher, J. C., Hobo, M., Meyerowitz, E. M. and Simon,R. (2000). "Dependence of stem cell fate in *Arabidopsis* on a feedback loop regulated by *CLV3* activity." *Science.* **289**: 617–619.

Broglie, K., Chet, I., Holliday, M., Crseeman, R., Biddle, P., Knowlton, S., Mauvasis, C.J. and Broglie, R. (1991). "Transgenic plants with enhanced resistance to the fungal pathogen *Rhizoctonia solani.*" *Science.* **254**: 1194–1197.

Carranco, R., Almoguera, C. and Jordano, J. (1997). "A plant small heat shock protein gene expressed during zygotic embryogenesis, but non-inducible by heat stress." *Journal of Biological Chemistry.* **272** (43): 27470–27475.

Constable, C.P., Bergery, D.R. and Ryan, C.A. (1995). "Systemin activates synthesis of wound-inducible tomato leaf polyphenol oxidase via the octadecanold defense signaling pathway." Proceedings of the National Academy of Science, USA. **92**: 407–411.

Doebley, J. and Lukens, L. (1998). "Transcriptional regulators and the evolution of plant form." *Plant Cell.* **10**: 1075–1082.

Ellis, J.G., Llewellyn, D.J., Walker, J.C., Dennis, E.S. and Peacock, W.J. (1987). "The ocs element: a 16 base pair palindrome essential for activity of the octopine synthase enhancer." *The EMBO Journal* **6**: 3203–3208.

Flint, L.H. (1936). "The action of radiation of specific wave-lengths in relation to the germination of light-sensitive lettuce seeds." Proc. Int. Seed Test.Assoc. **8**: 1–4.

Fraley, R.T., Rogers, S.G., Horsch, R.B., Sanders, P.R., Flick, J.S., Adams, S.P., Bittner, M.L., Brand, L.A., Fink, C.L., Fry, J.S., Galluppi, G.R., Goldberg, S.B., Hoffmann, N.L. and Woo, S.C. (1983). *PNAS*. **80**(15): 4803–4807.

Gan, S. and Amasino, R.M. (1995). "Inhibition of leaf senescence by auto-regulated production of cytokinin." *Science*. **270**: 1986–1988.

Glass, A.D.M. and Dunlop, J. (1974). "Influence of phenolic acids on ion uptake: IV. Depolarization of membrane potentials." *Plant Physiology*. **54**: 855–858.

Grebe, M., Gadea, G., Steinmann, T., Kientz, M., Rahfeld, J. U., Salchert, K., Koncz, C. and Jürgens, G. (2000). "A conserved domain of the *Arabidopsis* GNOM protein mediates subunit interaction and cyclophilin 5 binding." *Plant Cell*. **12**: 343–356.

Green, T. and Ryan, C. (1972). "Wound-induced proteinase inhibitor in plant leaves: a possible defense mechanism against insects." *Science*. **175**: 776–777.

Greiner, S., Rausch, T., Sonnewald, U. and Herbers, K. (1999). "Ectopic expression of a tobacco invertase inhibitor homolog prevents cold-induced sweetening of potato tubers." *Nature Biotechnology*. **17**: 708–711.

Grosset, J., Marty, I., Chartler, Y. and Meyer, Y. (1990). "mRNAs newly synthesized by tobacco mesophyll protoplasts are wound-inducible." *Plant Molecular Biology*. **15**: 485–496.

Gurevich, A.I., Tuzova, T.P., Shpak, E.D., Starkova, N.N., Esipov, R.S. and Miroshnikov, A.I. (1996). "Mechanism of action of the plant hormone Jasmonate I. Jasmonate-interacting proteins that regulate transcription of the pinII potato gene." *Bioorganicheskaya Khimiya*. **22**: 101–107.

Gurley, W.B. and Key, J.L. (1991). "Transcriptional regulation of the heat-shock response: a plant perspective." *Biochemistry*. **30**: 1–12.

Hagen, G., Kleinschmidt, A.J. and Guilfoyle, T.J. (1984). "Auxin-regulated gene expression in intact soybean hypocotyl and excised hypocotyls sections." *Planta.* **16**: 147–153.

Hake, S., Vollbrecht, E. and Freeling, M. (1989). "Cloning *KNOTTED*, the dominant morphological mutant in maize using Ds2 as a transposon tag." *EMBO J.* **8**: 15–22.

Hiatt, A.,Cafferkey, R. and Bowdish, K. (1989). "Production of antibodies in transgenic plants." *Nature.* **342**: 76–78.

Hildmann, T., Ebneth, M., Pena-Cortes, H., Sanchez-Serrano, J.J., Willmitzer, L. and Prat, S. (1992). "General roles of abscisic and jasmonic acids in gene activation as a result of mechanical wounding." *Plant Cell.* **4**: 1157–1170.

Holtorf, S., Apel, K. and Bohlmann, H. (1995). "Comparison of different Constitutive and inducible promoters for the overexpression of transgenes in *Arabidopsis thaliana*." *Plant Molecular Biology.* **29**: 637–646.

Ishimoto, M. and Crispeels, M.J. (1996). "Protective mechanism of the Mexican Bean Weevil against high levels of a a-amylase inhibitor in the common bean." *Plant Physiology.* **111**: 393–401.

Jarai, G., Truong, H.N., Daniel-Vedele, F. and Marzluf, G.A. (1992). "NIT2, the nitrogen regulatory protein of *Neurospora crassa*, binds upstream of *nia*, the tomato nitrate reductase gene, *in vitro*." *Current Genetics.* **21**: 37– 41.

Jin R.G., Liu, Y.B., Tabashnik B.E. and Borthakur, D. (2000). "Development of transgenic cabbage (*Brassica oleracea* var. *capitata*) for insect resistance by *Agrobacterium tumefaciens*-mediated transformation." *In vitro Cellular and Developmental Biology-Plant.* **36**: 231–237.

Kernan, A. and Thornburg, R.W. (1989). "Auxin levels regulate the expression of a wound-inducible proteinase inhibitor II-chloramphenicol acetyl transferase gene fusion *in vitro* and *in vivo*." *Plant Physiology.* **91**: 73–78.

Kim, S.R., Buckley, K., Costa, M.A. and An, G. (1994). "A 20 nucleotide upstream element is essential for the nopaline synthase (nos) promoter activity." *Plant Molecular Biology.* **24**: 105–117.

Kim, S.R., Kim, Y. and An, G. (1993). "Identification of methyljasmonate and salicylic acid response elements from nopaline synthase (nos) promoter." *Plant Physiology.* **103**: 97–103.

Krumm, T., Bandemer, K. and Boland, W. (1995). "Induction of volatile biosynthesis in the lima bean (*Phaseolus lunatus*) by leucine- and isoleucine conjugates of 1-oxo-and 1-hydroxyindan-4-carboxylic acid: evidence for amino acid conjugates of jasmonic acid as intermediates in the octadecanoid signaling pathway." FEBS Letters. **377**: 523–529.

Lee, J., Prathier, B. and Lobler, M. (1996). "Jasmonate signaling can be uncoupled from abscisic acid signaling in barley: identification of jasmonate-regulated transcripts which are not induced by abscisic acid." *Planta.* **199**: 625–632.

Letham, D.S. (1974). "Regulators of cell division in plant tissues XX. The cytokinins of coconut milk." *Physiologia Plantarum.* **32**: 66–70.

Liu, Z.B., Ulmasov, T., Shi, X., Hagen, G. and Guilfoyle, T.J. (1994). "The soybean GH3 promoter contains multiple auxin-inducible elements." *The Plant Cell.* **6**: 645–657.

Lloyd, A.M., Schena, M., Walbot, V. and Davis, R.W. (1994). "Epidermal cell fate determination in *Arabidopsis*: patterns defined by a steroid inducible regulator." *Science.* **266**: 436–439.

Lohman, K.N., Gan, S., John, M.C. and Amasino, R.M. (1994). "Molecular analysis of natural leaf senescence in *Arabidopsis thaliana*." *Physiologia Plantarum.* **92**: 322–328

Lucas, W. J., Bouche-Pillon, S., Jackson, D.P., Nguyen, L., Baker, L., Ding, B. and Hake, S. (1995). "Selective trafficking of KNOTTED1 homeodomain protein and its mRNA through plasmodesmata." *Science.* **270**: 1980–1983.

Lurin, C., Geelen, D., Barbier-Brygoo, H., Guern, J. and Maurel, C. (1996). "Cloning and functional expression of a plant voltage-dependent chloride channel." *Plant Cell.* **8**: 701–711.

Mason, H.S., Lam, D.M.K. and Amtzen, C.J. (1992). "Expression of hepatitis B surface antigen in transgenic plants." Proceedings of National Academy of Sciences, USA. **89**: 11745–11749.

Mauch, F., Mauch-Mani, B. and Boller, T. (1988). "Antifungal hydrolases in pea tissue: II. Inhibition of fungal growth by combinations of chitinase and b-1,3-glucanase." *Plant Physiology.* **88**: 936–942.

McGurl, B., Prarce, G., Orozco-Cardenas, M. and Ryan, C.A. (1992). "Structure, expression and antisense inhibition of the systemin precursor gene." *Science.* **255**: 1570–1573.

Mett, V.L., Podivinsky, E., Tennant, A.M., Lochhead, L.P., Jones, W.T. and Reynolds, P.H.S. (1996). "A system for tissue-specific copper controllable gene expression in transgenic plants: nodule-specific antisense of aspartate aminotranferase-P2." *Transgenic Research.* **5**: 105–113.

Milborrow, B.V. (2001). "The pathway of biosynthesis of abscisic acid in vascular plants: A review of the present state of knowledge of ABA biosynthesis." *Journal of Experimental Botany.* **52**:1145–1164.

Mohan, R., Peruman, V. and Kolattukudy, P.E. (1993). "Developmental and tissue-specific expression of a tomato anionic peroxidase gene by a minimal promoter, with wound and pathogen induction by an additional 5´-flanking region." *Plant Molecular Biology.* **22**: 475–490.

Mori, I.C. and Muto, S. (1997). "Abscisic acid activates a 48-kilodalton protein kinase in guard cell protoplasts." *Plant Physiology.* **113**: 833–839.

Muller-Rober, B.T., Sonnerwald, U. and Willmitzer, L. (1992). "Inhibition of the ADP-glucose pyrophosphorylase in transgenic potatoes leads to sugar-storing tubers and influences tuber formation and expression of tuber storage protein genes." *The EMBO Journal.* **11**(4): 1229–38.

Mullineaux, P.M. (1992). "Genetically engineered plants for herbicide resistance." In: *Plant Genetic Manipulation Crop Protection. Biotechnology in Agriculture Series.* Volume 7. Gatehouse, A.M.R., Hidder, V.A. and Boulder, D. Chap.7.

Mundy, J., Yamaguchi-Shinozak, K. and Chua, N.H. (1990). "Nuclear proteins bind conserved elements in the abscisic acid – responsive promoter of a rice *rab* gene." Proceedings of the National Academy of Sciences, USA. **87**: 1406–1410.

Murashige, T. and Skoog, F. (1962). "A revised medium for rapid growth and bioassays with tobacco tissue cultures." *Plant Physiol.* **15**(3): 473–497.

Nakai, A., Tanabe, M., Kawazoe, Y., Inazawa, J., Morimoto, R.I. and Nagata, K. (1997). "HSF4, a new member of the human heat shock factor family which lacks properties of a transcriptional activator." *Molecular Cell Biology.* **17**: 469–481.

Narvaez-Vassquez, J., Orozco-Cardenas, M.L. and Ryan, C.A. (1994). "Sulfhydryl reagent modulates systemic signaling for wound-induced and systemin-induced proteinase inhibitor synthesis." *Plant Physiology.* **105**: 725–730.

Neininger, A., Bichler, J., Schneiderbauer, A. and Mohr, H. (1993). "Response of a nitrite-reductase 3.1-kilobase upstream regulatory sequence from spinach to nitrate and light in transgenic tobacco. *Planta.* **189**: 440–442.

O'Donnell, P.J., Calvert, C., Atzorn, R., Wasternack, C., Leyser, H.M.O. and Bowles, D.J. (1996). "Ethylene as a signal mediating the wound response of tomato plants." *Science.* **274**: 1914–1917.

Pearce, G., Strydon, D., Johnson, S. and Ryan, C.A. (1991). "A polypeptide from tomato leaves induces wound-inducible proteinase inhibitor proteins." *Science.* **253**: 895–898.

Pen, J., Molendijk, L., Quax, W.J., Sijmons, P.C., van Ooyen, A.J.J., van den Elzen, P.J.M., Rietveld, K. and Hoekema, A. (1992). "Production of active *Bacillus licheniformis* alpha-amylase in tobacco and its application in starch liquefaction." *Biotechnology.* **10**: 292–296.

Pla, M., Gomez, J., Goday, A. and Pages, M. (1991). "Regulation of the abscisic acid-responsive gene *rab* 28 in maize viviparous mutants." *Molecular and General Genetics.* **230**: 394–400.

Premkumar,R., Sorger, G.J. and Gooden, D. (1980). "Repression of nitrate reductase in *Neurospora* studied by using L-methionine-DL-sulfoximine and glutamine auxotroph *gin-lb.*" *Journal of Bacteriology.* **143**: 411–415.

Ramputh, A.I. and Brown, A.W. (1996). "Rapid g-aminobutric acid synthesis and the inhibition of the growth and development of oblique-banded leaf-roller larvae." *Plant Physiology.* **111**: 1349–1352.

Rastogi, R., Back, E., Schneiderbauer, A., Bowsher, C.G., Moffatt, B. and Rothstein, S.J. (1993). "A 330 bp region of the spinach nitrite reductase gene promoter directs nitrate-inducible tissue-specific expression in transgenic tobacco." *Plant Journal.* **4**: 317–326.

Reinbothe, S., Reinbothe, C., Heintzen, C., Seidenbecher, C. and Parthier, B. (1993a). "A methyl jasmonate-induced shift in the length of the 5' untranslated region impairs transition of the plastid *rbcL* transcript in barley." *The EMBO Journal.* **12**: 15050–1512.

Rhodes, J.D., Thain, J.F. and Wildon, D.C. (1996). "The pathway for systemic electrical signal conduction in the wounded tomato plant." *Planta.* **200**: 50–57.

Rice, S.J., Grant, M.R., Reynolds, P.H.S. and Farnden, K.J.F. (1993). "DNA sequence of nodulin 45 from *Lupinus angustifolius.*" *Plant Science.* **90**: 155–166.

Riechmann, J.L. and Meyerowitz, E.M. (1997). "MADS domain proteins in plant development." *Biol. Chem.* **378**: 1079–1101.

Roeder, R.G. (1991). "The complexities of eukaryotic transcription initiation: regulation of preinitiation complex assembly." *Trends in Biochemical Sciences.* **16**: 402–408.

Ryan, C.A. (1981a). "Plant proteinases." In: *The Biochemistry of Plants: a Comprehensive Treatise*. Stumpf, P.K. and Conn, E.E. (eds.). Vol.6. Academic Press, New York, pp. 351–370.

Schaller, A., Bergey, D.R. and Ryan, C.A. (1995). "Induction of wound response genes in tomato leaves by bestatin, an inhibitor of amino-peptidases." *Plant Cell.* **7**: 1893–1898.

Severin, K. and Schoffl, F. (1990). "Heat inducible hygromycin resistance in transgenic tobacco." *Plant Molecular Biology.* **15**: 827–833.

Shen, Q. and Ho, T.H.D. (1995). "Functional dissection of an abscisic acid (ABA)-inducible gene reveals two independent ABA-responsive complexes each containing a G-box and a novel *cis*-acting element." *Plant Cell.* **7**: 295–307.

Shen, Q., Zhang, P. and Ho, T.H.D. (1996). "Modular nature of abscisic acid (ABA) response complexes: composite promoter units which are necessary and sufficient for ABA induction of gene expression in barley." *Plant Cell.* **8**: 1107–1119.

Shiu, S.H. and Bleecker, A.B. (2001). "Receptor-like kinases from *Arabidopsis* form a monophyletic gene family related to animal receptor kinases." Proc. Natl. Acad. Sci. USA. **98**: 10763–10768.

Singh, K., Tokuhisa, J.G., Dennis, E.S. and Peacock, W.J. (1989). "Saturation mutagenesis of the octopine synthase enhancer: correlation of mutant phenotypes with binding of a nuclear protein factor." Proceedings of the National Academy of Sciences, USA. **86**: 3733–3737.

Steinmann, T., Geldner, N., Grebe, M., Mangold, S. A., Jackson, C. L., Paris, S., Galweiler, L., Palme, K. and Jürgens, G. (1999). "Coordinated polar localization of auxin efflux carrier PIN1 by GNOM ARF GEF." *Science.* **286**: 316–318.

Strauch, E.W., Arnold, R., Alljah, W., Wohlleben, A., Puhler, P., Eckes, G., Donn, E., Ulmann, F. Hein and Wengenmayer, F. (1988). Eur.Patent EP 275957B1.

Thanavala, Y., Yang, Y.F., Lyons, P., Mason, H.S. and Amtzen, C.J. (1995). "Immunogenicity of transgenic plant-derived hepatitis B surface antigen. Proceedings of National Academy of Sciences, USA. **92**: 3358–3361.

Thornburg, R.W. and Li, X. (1991). "Wounding of the foliage of *Nicotiana tabacum* causes a decline in the levels of endogenous foliar IAA." *Plant Physiology.* **96**: 802–805.

Ulmasov, T., Hagen, G. and Guilfoyle, T.J. (1994). "The *ocs* element in the soybean GH2/4 promoter is activated by both active auxin and salicylic acid analogues." *Plant Molecular Biology.* **26**: 1055–1064.

Ulmasov, T., Hagen, G. and Guilfoyle, T.J. (1997a). "ARF1, a transcription factor that binds auxin response elements." *Science.* **276**: 1865–1868.

Ulmasov, T., Liu, Z.B., Hagen, G. and Guilfoyle, T.J. (1995b). "Composite structure of auxin response elements." *The Plant Cell.* **7**: 1611–1623.

Ulmasov, T., Murfett, J., Hagen, G. and Guilfoyle, T.J. (1997b). "Aux/ AA proteins repress expression of reporter genes containing natural and highly active synthetic auxin response elements." *Plant Cell.* **9**: 1963–1971.

Weinmann, P., Gossen, M., Hillen, W., Bujard, H. and Gatz, C. (1994). "A chimeric transactivator allows tetracycline-responsive gene expression in whole plants." *The Plant Journal.* **5**: 559–569.

Wildon, D.C., Thain, J.F., Minchin, P.E.H., Gubb, I.R., Reilly, A.M., Skipper, Y.D., Doherty, H.M., O'Donnell, P.J. and Bowles, D.J. (1992). "Electrical signaling and systemic proteinase inhibitor induction in the wounded plant." *Nature (London).* **360**: 62–65.

Wing, K.D. (1988). "RH5948, a nonsteroidal ecdysone agonist: effects on a *Drosophila* cell line." *Science.* **241**: 467–469.

Wray, J.L. (1993). "Molecular biology, genetics and regulation of nitrite reduction in higher plants." *Physiologia Plantarum.* **89**: 607–612.

Wright, C.F., Hamer, D.H. and McKenney, K. (1988). "Autoregulation of the yeast copper metallothionein gene depends on metal binding." *Journal of Biological Chemistry*. **263**: 1570–1574.

Zambryski, P. and Crawford, K. (2000). "Plasmodesmata: Gatekeepers for cell-to-cell transport of developmental signals in plants. *Annu. Rev. Cell. Dev. Biol.* **16**: 393–421.

Zimmerman, J.L., Apuya, N., Darwish, K. and O'Carroll, C. (1989). Novel regulation of heat shock genes during carrot somatic embryo development." *Plant Cell*. **1**: 1137–1146.

INDEX

Trichlorophenoxy acetic
    acid 271
Triple response 146
Trypsin inhibitor 306
Tryptophan 122
Type I diabetes 338

U

UDP glucose
    pyrophosphorylase 317
Ultraspiracle 223
Untranslated region 248
Upstream 14
Upstream activating
    sequences 225

V

Vascular cambium 166
Vicilins 207

Viral resistance 312
Viviparous 269
Vivipary 141

W

Wound-inducible gene 246

X

Xanthoxin 140

Y

Yeast galactose operon 225

Z

Zeatin 133
Zeaxanthin 139
Zein 203
Zinc fingers 222
Zona pellucida 338
Zygote 155

www.ingramcontent.com/pod-product-compliance
Lightning Source LLC
Chambersburg PA
CBHW070743160726
48004CB00001B/21